Metabolism of Bacterial Infection

Vijay Soni • Aditya Upadhyay
Editors

Metabolism of Bacterial Infection

Curated Literature Overview

Springer

Editors
Vijay Soni
Department of Medicine
Weill Cornell Medicine
New York, NY, USA

Aditya Upadhyay
Hospital for Special Surgery
New York, NY, USA

ISBN 978-3-031-46894-0 ISBN 978-3-031-46895-7 (eBook)
https://doi.org/10.1007/978-3-031-46895-7

This Springer imprint is published by the registered company Springer Nature Switzerland AG
The registered company address is: Gewerbestrasse 11, 6330 Cham, Switzerland

Preface

The metabolic landscape of bacterial infection is vast, complex, and increasingly recognized as central to understanding microbial pathogenesis and host responses. This book, *Metabolism of Bacterial Infection*, is born from the conviction that metabolism is not merely a supporting act in the biology of infection—it is a driving force that orchestrates virulence, immune evasion, drug resistance, and therapeutic outcomes. As the global burden of infectious diseases continues to evolve, so too must our approaches to studying and intervening in the metabolic processes that underpin them.

Over the past decade, advances in multi-omics technologies, computational modeling, and synthetic biology have illuminated the metabolic complexity of both pathogens and their hosts. This book brings together these insights in a structured and accessible format, offering a comprehensive view of how bacterial metabolism intersects with infection biology. The chapters in this volume are organized thematically to guide readers through the multifaceted roles of metabolism in bacterial pathogenesis. Each chapter integrates machine-curated summaries of peer-reviewed studies, offering readers a panoramic view of the field while maintaining scientific rigor.

The book begins with an overview of bacterial metabolism and its diverse roles in infection, setting the stage for deeper explorations into regulatory networks, nutrient acquisition, and metabolic signaling. Subsequent chapters explore the molecular mechanisms by which bacteria adapt to host environments, reprogram immune responses, and evolve resistance to antimicrobial agents. The book also highlights the systemic consequences of these interactions, including the impact on host immunometabolism and the microbiome. A significant portion of the book is dedicated to computational and mathematical modeling, reflecting the growing importance of systems biology in predicting metabolic behavior and identifying therapeutic targets. These models not only simulate bacterial physiology but also offer frameworks for designing novel interventions. The final chapters explore metabolic engineering and the development of next-generation therapeutics, showcasing how synthetic biology and natural product discovery are reshaping our ability to combat infectious diseases.

Throughout the book, emphasis is placed on integration—between experimental and computational approaches, between microbial and host metabolism, and between basic science and clinical relevance. The studies summarized here span a wide range of organisms, environments, and methodologies, yet they converge on a common theme: metabolism is a lens through which we can better understand and influence the course of bacterial infections.

This work is intended for microbiologists, clinicians, students, and researchers across disciplines who seek to deepen their understanding of infection biology through the prism of metabolism. It is both a reference and a call to action—to explore, innovate, and translate metabolic insights into meaningful solutions for global health.

I am deeply grateful to the researchers whose work forms the foundation of this book, and to the scientific community that continues to push the boundaries of what we know about microbial life. May this book serve as a resource, a guide, and an inspiration for those committed to unraveling the metabolic mysteries of bacterial infection.

New York, NY, USA

Vijay Soni
Aditya Upadhyay

About the Book

Auto-summaries can be generated by either an abstractive or an extractive auto-summarization:

- An extraction-based summarizer identifies the most important sentences of a text and uses the original sentences to create the summary.
- An abstraction-based summarizer creates new text based on deep learning. New phrases are created to summarize the content.

The auto-summaries you will find in this book have been generated via an abstractive summarization approach.

Each chapter was carefully edited by Dr. Vijay Soni and Dr. Aditya Upadhyay. The editors selected the papers, which were then auto-summarized. The editors have edited the auto-summaries to support the understanding of the original content. You will find the editors' reviews and guidance on the auto-summaries in the chapter introductions. In curated literature overview books, editors are defined as those who curate the content for the book by selecting the papers to be auto-summarized and by organizing the output into a meaningful order. Next to the thoughtful curation of the papers, editors should guide the readers through the auto-summaries and make transparent why they selected the papers. The ultimate goal is to provide a current literature review of Springer Nature publications on a given topic in order to support readers in overcoming information overload and to help them dive into a topic faster, to identify interdisciplinary overlaps, and to present papers that might not have been on the readers' radar.

Please note that the selected papers are not used to train an LLM while the auto-summaries are created.

Contents

About the Editors

Vijay Soni, Ph.D. has been affiliated with Weill Cornell Medicine, New York, since 2016, and brings over 15 years of research experience in microbial metabolism. He earned his Ph.D. in 2016 through a joint program between the National Institute of Immunology, New Delhi, and the Birla Institute of Technology and Science, Pilani, Hyderabad Campus, India. Following his doctoral studies, he joined Weill Cornell Medicine as a postdoctoral research associate and currently serves as an Instructor of Microbiology in Medicine.

His research focuses on the metabolism of the cell wall in *Mycobacterium tuberculosis*, with a particular emphasis on peptidoglycan remodeling and its implications for antibiotic resistance. By integrating metabolomics with multi-omics approaches, he investigates the complex metabolic reprogramming that occurs during antibiotic treatment and resistance development. His work aims to identify novel metabolic pathways, key metabolites, and regulatory mechanisms that can inform the development of new anti-tuberculosis therapies and combination treatments.

He is an active contributor to the scientific community, serving as an editor and reviewer for several high-impact journals. He has authored numerous peer-reviewed articles, reviews, editorials, book chapters, and books. His research has been recognized with multiple national and international awards. In addition to his academic pursuits, he is engaged in entrepreneurial ventures focused on drug discovery and diagnostic innovation.

Aditya Upadhyay, Ph.D. is a Postdoctoral Associate at the Hospital for Special Surgery, New York, where he investigates metabolic disorders, spine biology, and regenerative medicine. With over ten years in translational research, he has made impactful contributions to the understanding of fetal development in obesity and nonalcoholic fatty liver disease using humanized mouse models.

He earned his Ph.D. in molecular biology from SGPGIMS, India, in 2020, where he studied how maternal high-fat diets impair fetal organ development and how time-restricted feeding can mitigate these effects. He later joined Weill Cornell

Medicine (2021–2024), where his work on the PNPLA3 I148M variant advanced our knowledge of gene-diet interactions in liver disease.

Skilled in molecular biology, cell biology, histology, and humanized disease modeling, he has contributed to research on hepatocyte organoids, developmental epigenetics, and diabetic biomarkers. His work has appeared in journals like *BBA Molecular Basis of Disease* and *BBRC*, and he serves as a reviewer for *Frontiers*, *Cells*, and *PloS One*. He co-edited a 2025 special issue on metabolomics in infection.

He has earned several honors, including national fellowships and awards for scientific presentations. A passionate mentor, he trains students across educational levels and is dedicated to advancing precision medicine through collaborative, cross-disciplinary research.

Chapter 1
Introduction to Bacterial Metabolism and Different Infections

Vijay Soni and Aditya Upadhyay

1.1 Introduction by the Editor

Bacterial metabolism is a foundational aspect of microbial life, governing not only the biochemical processes essential for growth and survival but also the dynamic interactions between pathogens and their hosts. In the context of infectious diseases, metabolism acts as both a driver and a modulator of pathogenesis, manipulating virulence, immune evasion, and therapeutic outcomes. This chapter establishes the composite roles of bacterial metabolism in infection biology, portrayed on a variety of recent articles that collectively illustrate how metabolic pathways shape microbial behavior and host responses.

The chapter begins with highlights on the regulatory complexity of bacterial metabolism as exemplified *in Staphylococcus aureus*, where RNase III CLASH technology has revealed wide-ranging small RNA (sRNA) networks that couple metabolic conditions to toxin expression. These outcomes show that sRNA-mediated posttranscriptional regulation plays a significant role in modifying virulence in response to environmental signals, specifically under host-mimicking conditions. Such insights emphasize the importance of RNA-based regulatory systems in Gram-positive bacterial pathogens, which have historically been less understood compared to their Gram-negative counterparts.

Next, beyond individual pathogens, host metabolic responses to infection are gradually identified as central determinants of disease progression and resolution. In the context of allogeneic hematopoietic stem cell transplantation (HSCT), metabolomic profiling has shown substantial adaptations in host and microbiota-derived

V. Soni (✉)
Department of Medicine, Weill Cornell Medicine, New York, NY, USA
e-mail: vis2032@med.cornell.edu

A. Upadhyay
Hospital for Special Surgery, New York, NY, USA

V. Soni, A. Upadhyay (eds.), *Metabolism of Bacterial Infection*,
https://doi.org/10.1007/978-3-031-46895-7_1

metabolites linked with acute graft-versus-host disease (GvHD). These alterations, specifically in lipid and amino acid metabolism, indicate that metabolic signatures could serve as biomarkers for immune dysregulation and therapeutic stratification. The interaction between host metabolism and microbial communities thus emerges as a crucial axis in determining immune consequences.

Further, the concept of leveraging host metabolic pathways for therapeutic benefit is investigated through the study of specialized pro-resolving mediators (SPMs). Derived from essential fatty acids, SPMs such as resolvins and maresins actively promote the resolution of inflammation without compromising immune defense. Their ability to boost bacterial clearance and decrease antibiotic requirements emphasizes a promising avenue for adjunctive therapies that utilize endogenous resolution systems to stop infection and alleviate antibiotic resistance.

Interestingly, microbial metabolism also exerts systemic impact on host physiology, as shown by population-based studies relating gut microbiota composition to circulating metabolites. These studies uncover associations between specific microbial taxa and lipoprotein subclasses, indicating that microbial metabolic activity can affect host lipid profiles and cardiovascular risk. Such outcomes support the concept that microbial metabolism extends beyond localized infection sites, contributing to broader health consequences.

Additionally, structural innovations in bacterial metabolism are showcased by bacterial microcompartments (BMCs), protein-based organelles that encapsulate enzymes within selectively permeable shells. BMCs facilitate the compartmentalization of metabolic pathways, enhancing efficacy and shielding against toxic intermediates. Their modular architecture and functional diversity offer novel opportunities for bioengineering applications and synthetic biology.

Furthermore, in *Mycobacterium tuberculosis*, metabolic anticipation mechanisms facilitate the pathogen to persist hypoxic stress and promptly resume replication upon re-aeration. This defensive response involves the strategic accumulation and mobilization of metabolic intermediates, positioning the bacterium for effective cell wall biogenesis and immune modulation. Such adaptations emphasize the evolutionary sophistication of persistent pathogens in directing hostile host environments.

Finally, methodological developments, such as metabolic labeling of peptidoglycan carbohydrates, have opened new opportunities for visualizing bacterial cell wall dynamics and immune interactions. These techniques provide a high-resolution understanding of peptidoglycan architecture and its role in host–pathogen communication. Additionally, the metabolism of glycine, serine, and threonine has been demonstrated to affect complement-mediated killing, indicating that metabolic reprogramming could reinstate immune efficacy against serum-resistant bacteria.

Collectively, the studies presented in this chapter demonstrate that bacterial metabolism is not a passive backdrop to infection but an active and adaptive force that shapes the trajectory of disease. By integrating molecular, cellular, and systemic perspectives, this chapter lays the groundwork for deeper explorations into the metabolic underpinnings of bacterial pathogenesis and the potential for metabolism-informed interventions in infectious disease management.

1.2 Machine Generated Summaries

Disclaimer: The summaries in this chapter were generated from Springer Nature publications using abstractive AI auto-summarization: An abstraction-based summarizer creates new text based on deep learning. New phrases are created to summarize the content. As the constituted sentences are machine selected, they may not fully reflect the body of the work, so we strongly advise that the original content is read and cited. The auto-generated summaries were curated by the editor to meet Springer Nature publication standards.

RNase III CLASH in MRSA Uncovers sRNA Regulatory Networks Coupling Metabolism to Toxin Expression [95]

This is a machine-generated summary of:

McKellar, Stuart W.; Ivanova, Ivayla; Arede, Pedro; Zapf, Rachel L.; Mercier, Noémie; Chu, Liang-Cui; Mediati, Daniel G.; Pickering, Amy C.; Briaud, Paul; Foster, Robert G.; Kudla, Grzegorz; Fitzgerald, J. Ross; Caldelari, Isabelle; Carroll, Ronan K.; Tree, Jai J.; Granneman, Sander RNase III CLASH in MRSA Uncovers sRNA Regulatory Networks Coupling Metabolism to Toxin Expression [95]

Published in: Nature Communications (2022)

Link to original: https://doi.org/10.1038/s41467-022-31173-y

If you want to cite the papers, please refer to the original.

For technical reasons we could not place the page where the original quote is coming from.

Research Highlights

Bacteria, such as *Staphylococcus aureus*, adapt to stress through dynamic transcriptional networks and posttranscriptional regulatory mechanisms involving small RNAs (sRNAs). While sRNAs in Gram-negative bacteria like *Escherichia coli* are well-studied, their roles in Gram-positive species remain less understood. Now, a new study by Stuart W. McKellar and colleagues explores the function of sRNAs in *S. aureus*, revealing novel insights into their regulatory roles during host infection.

The study employed the Cross-linking, Ligation And Sequencing of Hybrids (CLASH) technique to identify sRNA-RNA interactions in *S. aureus* under conditions mimicking the host environment. By using RNase III as a bait protein, the researchers enriched for sRNA-RNA duplexes and discovered hundreds of novel interactions, many of which were condition-specific. The study also identified numerous mRNA-mRNA and sRNA-sRNA interactions, suggesting that metabolic pathways are interconnected through RNA-RNA interactions. The findings highlight the extensive role of RNase III in sRNA-mediated regulation and reveal that toxin expression in *S. aureus* is subject to significant sRNA-mediated control.

The research significantly expands the known repertoire of sRNA-target interactions in *S. aureus* and provides insights into how these interactions contribute to virulence regulation in response to metabolic changes. Future study could explore the mechanistic details of these interactions and their implications for bacterial adaptation during infection. The identification of RNase III as a key player in these processes suggests potential avenues for therapeutic intervention targeting bacterial virulence mechanisms.

Metabolomics Analysis of Human Acute Graft-Versus-Host Disease Reveals Changes in Host and Microbiota-Derived Metabolites [96]

This is a machine-generated summary of:

Michonneau, David; Latis, Eleonora; Curis, Emmanuel; Dubouchet, Laetitia; Ramamoorthy, Sivapriya; Ingram, Brian; de Latour, Régis Peffault; Robin, Marie; de Fontbrune, Flore Sicre; Chevret, Sylvie; Rogge, Lars; Socié, Gérard Metabolomics Analysis of Human Acute Graft-Versus-Host Disease Reveals Changes in Host and Microbiota-Derived Metabolites [96]

Published in: Nature Communications (2019)

Link to original: https://doi.org/10.1038/s41467-019-13498-3

If you want to cite the papers, please refer to the original.

For technical reasons we could not place the page where the original quote is coming from.

Research Highlights

Allogeneic hematopoietic stem cell transplantation (HSCT) is a critical treatment for hematologic malignancies and various hematopoietic disorders. Despite its therapeutic potential, HSCT is often complicated by graft-versus-host disease (GvHD), which significantly contributes to patient morbidity and mortality. While previous research has primarily focused on the role of T cell subsets in GvHD pathophysiology, recent studies suggest that external factors such as tissue microenvironment and host microbiota may also influence the allogeneic immune response. Now, a new study by David Michonneau and colleagues explores the impact of metabolomics on HSCT outcomes, revealing significant changes in metabolomics profiles associated with acute GvHD onset.

The study analyzed two cohorts of patients who underwent allogeneic-HSCT from HLA-identical sibling donors. Using high-throughput metabolomics, the researchers examined plasma samples from both donors and recipients to identify changes in circulating metabolites. The analysis employed ultrahigh performance liquid chromatography-tandem mass spectrometry to detect metabolites, focusing on variations between recipients with and without GvHD. The study found that HSCT recipients exhibited significant alterations in their metabolome compared to healthy donors, particularly in lipid and amino acid metabolism. Notably, tryptophan-derived metabolites were decreased in recipients without GvHD, while polyamine metabolites increased. In contrast, recipients with acute GvHD showed distinct metabolomic changes, including reduced detection of indolepropionate and decreased levels of amino acids and lipids.

The findings suggest that metabolomic alterations following HSCT may play a role in modulating immune responses and influencing GvHD onset. The study highlights the potential of using metabolomics profiles as biomarkers for predicting GvHD risk and tailoring therapeutic interventions. Future studies could explore the functional consequences of these metabolomic changes, particularly the role of microbiota-derived metabolites in regulating immune responses during GvHD. These insights could pave the way for novel strategies to mitigate GvHD severity and improve transplantation outcomes by targeting specific metabolic pathways involved in immune regulation.

Infection Regulates Pro-Resolving Mediators that Lower Antibiotic Requirements [97]

This is a machine-generated summary of:

Chiang, Nan; Fredman, Gabrielle; Bäckhed, Fredrik; Oh, Sungwhan F.; Vickery, Thad; Schmidt, Birgitta A.; Serhan, Charles N. Infection regulates pro-resolving mediators that lower antibiotic requirements [97]

Published in: Nature (2012)

Link to original: https://doi.org/10.1038/nature11042

If you want to cite the papers, please refer to the original.

For technical reasons we could not place the page where the original quote is coming from.

Research Highlights

The acute inflammatory response is a crucial defense mechanism evolved to eliminate invading organisms, ideally resolving itself to restore homeostasis. Recent evidence suggests that the resolution of acute inflammation is an active process involving the biosynthesis of specialized pro-resolving mediators (SPMs) derived from essential fatty acids. These mediators, such as lipoxins and resolvins, play a significant role in controlling inflammation without causing immunosuppression and are protective in various organs. However, the specific roles of SPMs in resolving bacterial infections remain unclear. Now, a new study by Nan Chiang and colleagues investigates the involvement of SPMs in resolving *E. coli* infections using a murine peritonitis model.

The study utilized a well-established murine model of peritonitis to explore how different doses of *E. coli* affect the resolution of infection and the role of SPMs in this process. Mice were inoculated with either a low or high dose of *E. coli*, and the inflammatory response was monitored over time. The researchers employed mass-spectrometry-based lipidomics to profile local lipid mediators in exudates from these infections. They found that lower doses of *E. coli* led to self-limited infections with increased levels of pro-resolving mediators like maresins and resolvins, while higher doses resulted in prolonged inflammation with elevated pro-inflammatory leukotrienes. The study also demonstrated that SPMs enhanced phagocyte containment and clearance of bacteria, reduced hypothermia, and increased survival rates in infected mice.

This research advances our understanding of how SPMs contribute to resolving bacterial infections and highlights their potential therapeutic applications. The findings suggest that SPMs can enhance antibiotic effectiveness, as demonstrated by their ability to improve outcomes when combined with ciprofloxacin or vancomycin in treating *E. coli* and *S. aureus* infections, respectively. Future studies could explore the clinical implications of these findings, particularly in developing strategies to reduce antibiotic use by targeting host resolution pathways alongside traditional antibacterial treatments. This approach may offer new avenues for addressing antibiotic resistance by leveraging the body's natural resolution mechanisms.

Relationship Between Gut Microbiota and Circulating Metabolites in Population-Based Cohorts [98]

This is a machine-generated summary of:
Vojinovic, Dina; Radjabzadeh, Djawad; Kurilshikov, Alexander; Amin, Najaf; Wijmenga, Cisca; Franke, Lude; Ikram, M. Arfan; Uitterlinden, Andre G.; Zhernakova, Alexandra; Fu, Jingyuan; Kraaij, Robert; van Duijn, Cornelia M. Relationship between gut microbiota and circulating metabolites in population-based cohorts [98]
Published in: Nature Communications (2019)
Link to original: https://doi.org/10.1038/s41467-019-13721-1

Research Highlights

The gut microbiota has garnered significant attention for its role in major human diseases, particularly through its metabolic and immune signals that enter the circulation. It has been linked to conditions such as obesity, diabetes, and dyslipidemia, with recent studies highlighting its influence on circulating lipid levels like triglycerides and HDL cholesterol. However, the specific associations between gut microbiota and detailed lipoprotein subclasses remain unclear. Now, a new study by Dina Vojinovic and colleagues provides novel insights into these relationships by examining the metabolome using nuclear magnetic resonance technology in two large population-based cohorts: the Rotterdam Study and LifeLines-DEEP.

The study involved a comprehensive analysis of gut microbiota composition and circulating metabolites in participants from these cohorts. Researchers used H-NMR technology to quantify a wide range of metabolites in fasting plasma samples, including amino acids, glycolysis-related metabolites, ketone bodies, fatty acids, routine lipids, and lipoprotein subclasses. Fecal samples were collected to profile gut microbiota using rRNA sequencing. The analysis revealed significant associations between specific microbial families and genera with various lipoprotein particles such as VLDL and HDL subfractions. For instance, family Christensenellaceae was linked to VLDL particles of various sizes and small HDL particles. The study also confirmed associations between serum triglycerides and genus *Ruminococcus gnavus*. These findings were adjusted for potential confounders like age, sex, BMI, medication use, smoking status, and alcohol consumption.

The study advances our understanding of how gut microbiota influences host lipid levels by identifying specific microbial taxa associated with distinct lipoprotein subclasses. This highlights the potential of targeting gut microbiota for therapeutic interventions in metabolic diseases. Future study could focus on exploring metabolites detected by other techniques beyond the Nightingale platform to capture a broader range of circulating compounds. Additionally, longitudinal studies collecting stool and blood samples over time could provide insights into dynamic changes in gut microbiota and metabolite interactions. The findings underscore the importance of considering microbial diversity in relation to health outcomes such as cardiovascular disease risk.

Bacterial Microcompartments [99]

This is a machine-generated summary of:

Kerfeld, Cheryl A.; Aussignargues, Clement; Zarzycki, Jan; Cai, Fei; Sutter, Markus Bacterial microcompartments [99]

Published in: Nature Reviews Microbiology (2018)

Link to original: https://doi.org/10.1038/nrmicro.2018.10

If you want to cite the papers, please refer to the original.

For technical reasons we could not place the page where the original quote is coming from.

Research Highlights

Bacterial microcompartments (BMCs) are protein-based organelles in bacteria that function similarly to eukaryotic organelles, encapsulating enzymes within a selectively permeable protein shell. This structure allows BMCs to sequester metabolic pathways, protecting the cell from toxic intermediates and enhancing metabolic efficiency. While the architecture of BMC shells is conserved, their functions vary widely, participating in both anabolic and catabolic processes. Despite advances in understanding BMC structure and function, there remains a gap in knowledge regarding the diversity of BMC types and their assembly mechanisms. Now, a new study by Cheryl A. Kerfeld and colleagues explores recent discoveries in BMC biology, highlighting the structural conservation of shell proteins and the potential for bioengineering applications.

The study delves into the structural components of BMC shells, which are composed of three types of proteins: BMC-H, BMC-T, and BMC-P. These proteins form hexamers, trimers, and pentamers that assemble into polyhedral shells with pores allowing metabolite passage. The research outlines how these structures are conserved across different bacterial species despite functional diversity. The authors describe the assembly pathways of various BMCs, such as carboxysomes and metabolosomes, noting differences in how their cores and shells form. Techniques like bioinformatics and advanced microscopy have facilitated the identification of new BMC types across bacterial phyla, expanding our understanding of their roles in metabolism and potential applications in synthetic biology.

The advances outlined by Kerfeld et al. pave the way for future studies to explore the engineering of BMCs for novel functions. By leveraging the modular nature of BMC gene clusters, researchers can potentially introduce these metabolic modules into different organisms to enhance or create new metabolic pathways. Future studies could focus on characterizing unclassified BMCs and exploring their roles in microbial ecology and pathogenesis. Additionally, understanding how to manipulate shell permeability and core enzyme interactions could lead to innovative applications in biotechnology and medicine. The study underscores the potential for BMCs to serve as versatile tools in bioengineering, offering insights into both fundamental biology and practical applications.

Metabolic Anticipation in **Mycobacterium tuberculosis** *[100]*

This is a machine-generated summary of:

Eoh, Hyungjin; Wang, Zhe; Layre, Emilie; Rath, Poonam; Morris, Roxanne; Branch Moody, D.; Rhee, Kyu Y. Metabolic anticipation in *Mycobacterium tuberculosis* [100]

Published in: Nature Microbiology (2017)

Link to original: https://doi.org/10.1038/nmicrobiol.2017.84

If you want to cite the papers, please refer to the original.

For technical reasons we could not place the page where the original quote is coming from.

Research Highlights
Mycobacterium tuberculosis, the bacterium responsible for tuberculosis, is a significant global health threat, with humans as its primary reservoir. The bacterium's ability to enter a dormant state and later reactivate is crucial for its survival and transmission. Hypoxia, or low oxygen levels, is a common condition encountered by *M. tuberculosis* within host environments, which halts its replication and induces drug tolerance. However, when oxygen levels are restored, the bacterium can resume replication. Despite extensive research on *M. tuberculosis*'s response to hypoxia, the mechanisms that enable it to survive and recover from such conditions remain unclear. Now, a new study by Hyungjin Eoh and colleagues reveals a distinct set of metabolic adaptations in *M. tuberculosis* that facilitate survival through hypoxia and recovery upon re-aeration.

The study utilized an in vitro system to simulate hypoxic conditions and observed metabolic changes in *M. tuberculosis*. Researchers found that hypoxia triggered the accumulation of specific metabolites in glycolysis, the pentose phosphate pathway, and aminosugar biosynthesis. These changes were reversible upon re-aeration and were linked to decreases in certain glycolytic intermediates while maintaining glycogen stores. The team discovered that these accumulations predominantly occurred from preexisting metabolic stores rather than newly synthesized metabolites. Further investigation revealed that trehalose mycolates on the bacterial cell surface were catabolized during hypoxia, leading to changes in immunostimulatory lipid levels and reduced cytokine induction in macrophages. The study also identified the TreS enzyme as essential for these metabolic responses, as its deletion impaired the bacterium's ability to adapt to hypoxia.

The findings suggest that *M. tuberculosis* has evolved an anticipatory metabolic response to hypoxia that prepares it for subsequent oxygen exposure by poising it for peptidoglycan biosynthesis upon re-aeration. This mechanism may help the bacterium overcome thermodynamic barriers during cell cycle reentry after dormancy. Future study could explore how these metabolic adaptations contribute to *M. tuberculosis*'s pathogenicity and persistence within host environments over extended periods. Understanding these processes may provide new insights into potential therapeutic targets for combating tuberculosis by disrupting these anticipatory responses or their regulatory pathways.

Metabolic Labeling of the Carbohydrate Core in Bacterial Peptidoglycan and Its Applications [101]

This is a machine-generated summary of:

Liang, Hai; DeMeester, Kristen E.; Hou, Ching-Wen; Parent, Michelle A.; Caplan, Jeffrey L.; Grimes, Catherine L. Metabolic labeling of the carbohydrate core in bacterial peptidoglycan and its applications [101]

Published in: Nature Communications (2017)
Link to original: https://doi.org/10.1038/ncomms15015

If you want to cite the papers, please refer to the original.

For technical reasons we could not place the page where the original quote is coming from.

Research Highlights

Peptidoglycan (PG) is a crucial component of bacterial cell walls, playing a vital role in cell division, shape maintenance, and immune system recognition. The innate immune system detects PG fragments, which can influence the development of diseases like Crohn's disease and asthma. However, understanding the biological identity and generation of these fragments is challenging due to limited tools for tracking NAM-containing PG fragments. Now, a new study by Hai Liang and colleagues addresses this gap by developing a method to synthesize NAM derivatives and incorporate them into bacterial PG for visualization.

The study utilized bioorthogonal chemistry to modify the NAM carbohydrate backbone of PG. By leveraging the promiscuity of bacterial recycling enzymes AmgK and MurU, the researchers synthesized NAM derivatives with bioorthogonal modifications. These modified NAMs were incorporated into the PG of *E. coli* and other bacteria using metabolic labeling techniques. The labeled bacteria were then visualized using advanced microscopy techniques such as structured illumination microscopy (SIM) and stochastic optical reconstruction microscopy (STORM). This approach allowed for the direct visualization of PG carbohydrates in bacterial cells, providing insights into their distribution during cell division.

The development of this NAM-based labeling method represents a significant advancement in studying bacterial cell walls. It enables detailed visualization of PG architecture and offers potential applications in tracking bacterial interactions with immune cells. Future study could explore the use of these labeled bacteria to investigate immune responses to specific PG fragments or to further understand the coordination between PG synthesis and bacterial cell division. This method could also be adapted to study a broader range of bacterial species due to the conserved nature of PG biosynthesis across different bacteria.

Glycine, Serine and Threonine Metabolism Confounds Efficacy of Complement-Mediated Killing [102]

This is a machine-generated summary of:

Cheng, Zhi-xue; Guo, Chang; Chen, Zhuang-gui; Yang, Tian-ci; Zhang, Jian-ying; Wang, Jie; Zhu, Jia-xin; Li, Dan; Zhang, Tian-tuo; Li, Hui; Peng, Bo; Peng, Xuan-xian Glycine, Serine and ThreonineMetabolism Confounds Efficacy of Complement-Mediated Killing [102]

Published in: Nature Communications (2019)

Link to original: https://doi.org/10.1038/s41467-019-11129-5

Research Highlights

The human complement system plays a crucial role in defending against bacterial infections by killing bacteria in the bloodstream and on mucosal surfaces. However, some bacteria, known as serum-resistant bacteria, can survive and proliferate despite a fully functional complement system, leading to serious systemic infections. While previous studies have focused on the role of bacterial outer membrane proteins and lipopolysaccharides in serum resistance, the metabolic state of these bacteria remains poorly understood. Now, a new study by Zhi-xue Cheng and colleagues explores the metabolic pathways involved in serum resistance, revealing that glycine catabolism is significantly reduced in serum-surviving *E. coli*.

To investigate this phenomenon, the researchers conducted a comprehensive metabolomic profiling of *E. coli* exposed to serum. They found that metabolites involved in glycine, serine, and threonine metabolism were present at lower concentrations in serum-surviving cells compared to controls. Using qRT-PCR, they demonstrated that the expression of genes mediating glycine flux to the TCA cycle was decreased under serum exposure. The study also showed that adding exogenous glycine, serine, or threonine could restore sensitivity to serum killing by altering the metabolic state of the bacteria. This was further supported by experiments showing increased membrane potential and NADH production when glycine was added to serum-exposed cells.

The findings suggest that targeting metabolic pathways could be a novel strategy for combating serum-resistant bacterial infections. By reprogramming the metabolome with exogenous metabolites like glycine, it may be possible to enhance the bactericidal activity of serum against resistant strains. Future studies could explore the potential of this approach in clinical settings and its applicability to other multidrug-resistant pathogens. Additionally, understanding how these metabolic changes interact with other cellular processes could provide further insights into overcoming bacterial resistance mechanisms.

1.3 Conclusion

The exploration of bacterial metabolism through the lens of infection biology reveals a landscape of remarkable complexity, adaptability, and clinical relevance. As demonstrated by the diverse studies summarized in this chapter, metabolism is not merely a background process in microbial life—it is a central axis around which bacterial survival, virulence, and host interactions revolve. This chapter has highlighted how metabolic pathways serve as both sensors and effectors in the infectious process, shaping the trajectory of disease and proposing novel insights for therapeutic intervention.

One of the most striking themes emerging from these studies is the regulatory sophistication of bacterial metabolism. In *Staphylococcus aureus*, the discovery of extensive sRNA-mediated networks that link metabolic states to toxin expression highlights the significance of posttranscriptional control in virulence regulation. These findings imply that metabolic signals are profoundly rooted in the bacterial decision-making architecture, allowing pathogens to fine-tune their behaviors in response to environmental stress and host defenses.

Similarly compelling is the role of host metabolism in modifying infection consequences. The metabolomic profiling of patients undergoing hematopoietic stem cell transplantation shows that alterations in host and microbiota-derived metabolites are closely correlated with the beginning of graft-versus-host disease. These insights indicate the potential of metabolic biomarkers in predicting immune dysregulation and tailoring clinical interventions. Moreover, the identification of specific pro-resolving mediators (SPMs) as key players in infection resolution presents a paradigm shift in therapeutic approach—one that accentuates enhancing the body's natural resolution pathways rather than exclusively relying on antimicrobial agents.

The systemic influence of microbial metabolism is further explained by population-based studies connecting gut microbiota composition to circulating lipid profiles. These associations suggest that microbial communities exert far-reaching effects on host physiology, with implications for metabolic diseases and cardiovascular risk. Such findings strengthen the need to study microbial metabolism not only in the context of localized infections but also as a contributor to broader health outcomes.

Structural innovations in bacterial metabolism, such as bacterial microcompartments (BMCs), add another level of complexity to the metabolic landscape. These protein-based organelles enable the compartmentalization of metabolic reactions, increasing effectiveness and safeguarding against toxic intermediates. Their modular architecture and evolutionary conservation across bacterial species emphasize their prospective as tools for synthetic biology and metabolic engineering.

In *Mycobacterium tuberculosis*, the concept of metabolic anticipation presents a powerful model of how pathogens prepare for environmental changes. The ability of *M. tuberculosis* to accumulate and mobilize metabolic intermediates during hypoxia and re-aeration reflects a strategic adaptation that advocates persistence and reactivation. These findings expand our understanding of dormancy and drug tolerance in chronic infections and propose new targets for interrupting bacterial endurance mechanisms.

Technological advances, such as metabolic labeling of peptidoglycan carbohydrates, have opened new frontlines in visualizing bacterial cell wall dynamics and immune interactions. These tools provide high-resolution insights into the spatial organization of bacterial metabolism and its role in host–pathogen communication. Additionally, the study of amino acid metabolism—specifically glycine, serine, and threonine—reveals that metabolic reprogramming can influence complement-mediated killing, presenting a novel approach to kill serum resistance in pathogenic bacteria.

Taken together, the studies in this chapter demonstrate that bacterial metabolism is a dynamic and integrative force in infection biology. It governs not only the internal logic of microbial cells but also their interactions with the host and the environment. As we move forward in this book, the foundational concepts and discoveries outlined here will inform deeper explorations into specific metabolic pathways, host–microbe interactions, and translational applications. Understanding and manipulating bacterial metabolism holds enormous promise for developing both basic science and clinical practice, specifically in the face of soaring antimicrobial resistance and the need for innovative therapeutic tactics.

Bibliography

1. Ivain, L. et al. An in vivo reporter assay for sRNA-directed gene control in Gram-positive bacteria: Identifying a novel sRNA target in Staphylococcus aureus. *Nucleic Acids Res. 45*, 4994–5007 (2017).
2. Liu, W. et al. Assessment of bona fide sRNAs in Staphylococcus aureus. *Front. Microbiol. 9*, 228 (2018).

3. Pinel-Marie, M. L., Brielle, R. & Felden, B. Dual Toxic-Peptide-Coding Staphylococcus aureus RNA under Antisense Regulation Targets Host Cells and Bacterial Rivals Unequally. *Cell Rep. 7*, 424–435 (2014).
4. Sayed, N., Jousselin, A. & Felden, B. A cis-antisense RNA acts in trans in Staphylococcus aureus to control translation of a human cytolytic peptide. *Nat. Struct. Mol. Biol. 19*, 105–113 (2012).
5. Guillet, J., Hallier, M. & Felden, B. Emerging Functions for the Staphylococcus aureus RNome. *PLoS Pathog. 9*, 1–13 (2013).
6. Lioliou, E. et al. Global regulatory functions of the Staphylococcus aureus endoribonuclease III in gene expression. *PLoS Genet 8*, e1002782 (2012).
7. Mellor, A. L. & Munn, D. H. Ido expression by dendritic cells: tolerance and tryptophan catabolism. *Nat. Rev. Immunol. 4*, 762–774 (2004).
8. Shono, Y. et al. Increased GVHD-related mortality with broad-spectrum antibiotic use after allogeneic hematopoietic stem cell transplantation in human patients and mice. *Sci. Transl. Med. 8*, 339ra71 (2016).
9. Peled, J. U. et al. Intestinal microbiota and relapse after hematopoietic-cell transplantation. *J. Clin. Oncol. 35*, 1650–1659 (2017).
10. Mathewson, N. D. et al. Gut microbiome-derived metabolites modulate intestinal epithelial cell damage and mitigate graft-versus-host disease. *Nat. Immunol. 17*, 505–513 (2016).
11. Fujiwara, H. et al. Microbial metabolite sensor GPR43 controls severity of experimental GVHD. *Nat. Commun. 9*, 3674 (2018).
12. Scott, N. A. et al. Antibiotics induce sustained dysregulation of intestinal T cell immunity by perturbing macrophage homeostasis. *Sci. Transl. Med. 10*, eaao4755 (2018).
13. Riwes, M. & Reddy, P. Microbial metabolites and graft versus host disease. *Am. J. Transplant. 18*, 23–29 (2018).
14. Serhan, C. N. Resolution phases of inflammation: novel endogenous anti-inflammatory and pro-resolving lipid mediators and pathways. *Annu. Rev. Immunol. 25*, 101–137 (2007)
15. De Caterina, R. n−3 fatty acids in cardiovascular disease. *N. Engl. J. Med. 364*, 2439–2450 (2011)
16. Yang, R., Chiang, N., Oh, S. F. & Serhan, C. N. Metabolomics-lipidomics of eicosanoids and docosanoids generated by phagocytes. *Curr. Protoc. Immunol. 95*, 14.26.11–14.26.26 (2011)
17. Komaroff, A. L. The microbiome and risk for obesity and diabetes. *J. Am. Med. Assoc. 317*, 355–356 (2017).
18. Zhernakova, A. et al. Population-based metagenomics analysis reveals markers for gut microbiome composition and diversity. *Science 352*, 565–569 (2016).
19. Fu, J. et al. The gut microbiome contributes to a substantial proportion of the variation in blood lipids. *Circ. Res. 117*, 817–824 (2015).
20. Rothschild, D. et al. Environment dominates over host genetics in shaping human gut microbiota. *Nature 555*, 210 (2018).
21. Zierer, J. et al. The fecal metabolome as a functional readout of the gut microbiome. *Nat. Genet. 50*, 790–+ (2018).
22. Holmes, M. V. et al. Lipids, lipoproteins, and metabolites and risk of myocardial infarction and stroke. *J. Am. Coll. Cardiol. 71*, 620–632 (2018).
23. Wurtz, P. et al. Metabolite profiling and cardiovascular event risk: a prospective study of 3 population-based cohorts. *Circulation 131*, 774–785 (2015).
24. Nakamura, H. et al. Plasma amino acid profiles are associated with insulin, C-peptide and adiponectin levels in type 2 diabetic patients. *Nutr. Diabetes 4*, e133 (2014).
25. Magnusson, M. et al. A diabetes-predictive amino acid score and future cardiovascular disease. *Eur. Heart J. 34*, 1982–1989 (2013).
26. Le Chatelier, E. et al. Richness of human gut microbiome correlates with metabolic markers. *Nature 500*, 541–546 (2013).

27. Biddle, A., Stewart, L., Blanchard, J. & Leschine, S. Untangling the genetic basis of fibrolytic specialization by Lachnospiraceae and Ruminococcaceae in diverse gut communities. *Diversity 5*, 627–640 (2013).
28. Kakiyama, G. et al. Modulation of the fecal bile acid profile by gut microbiota in cirrhosis. *J. Hepatol. 58*, 949–955 (2013).
29. Bonder, M. J. et al. The effect of host genetics on the gut microbiome. *Nat. Genet. 48*, 1407–1412 (2016).
30. Kerfeld, C. A., Heinhorst, S. & Cannon, G. C. Bacterial microcompartments. *Annu. Rev. Microbiol. 64*, 391–408 (2010).
31. Kerfeld, C. A. & Melnicki, M. R. Assembly, function and evolution of cyanobacterial carboxysomes. *Curr. Opin. Plant Biol. 31*, 66–75 (2016).
32. Rae, B. D., Long, B. M., Badger, M. R. & Price, G. D. Functions, compositions, and evolution of the two types of carboxysomes: polyhedral microcompartments that facilitate CO2 fixation in cyanobacteria and some proteobacteria. *Microbiol. Mol. Biol. Rev. 77*, 357–379 (2013).
33. Turmo, A., Gonzalez-Esquer, C. R. & Kerfeld, C. A. Carboxysomes: metabolic modules for CO2 fixation. *FEMS Microbiol. Lett. 364*, fnx176 (2017).
34. Chowdhury, C., Sinha, S., Chun, S., Yeates, T. O. & Bobik, T. A. Diverse bacterial microcompartment organelles. *Microbiol. Mol. Biol. Rev. 78*, 438–468 (2014).
35. Axen, S. D., Erbilgin, O. & Kerfeld, C. A. A taxonomy of bacterial microcompartment loci constructed by a novel scoring method. *PLOS Comput. Biol. 10*, e1003898 (2014). *This is a comprehensive survey of BMC types and subtypes by bioinformatic analysis of superloci.*
36. Kerfeld, C. A. & Erbilgin, O. Bacterial microcompartments and the modular construction of microbial metabolism. *Trends Microbiol. 23*, 22–34 (2015).
37. Zarzycki, J., Erbilgin, O. & Kerfeld, C. A. Bioinformatic characterization of glycyl radical enzyme-associated bacterial microcompartments. *Appl. Environ. Microbiol. 81*, 8315–8329 (2015).
38. Yeates, T. O., Kerfeld, C. A., Heinhorst, S., Cannon, G. C. & Shively, J. M. Protein-based organelles in bacteria: carboxysomes and related microcompartments. *Nat. Rev. Microbiol. 6*, 681–691 (2008).
39. Frank, S., Lawrence, A. D., Prentice, M. B. & Warren, M. J. Bacterial microcompartments moving into a synthetic biological world. *J. Biotechnol. 163*, 273–279 (2013).
40. Gonzalez-Esquer, C. R., Newnham, S. E. & Kerfeld, C. A. Bacterial microcompartments as metabolic modules for plant synthetic biology. *Plant J. 87*, 66–75 (2016).
41. Plegaria, J. S. & Kerfeld, C. A. Engineering nanoreactors using bacterial microcompartment architectures. *Curr. Opin. Biotechnol. 51*, 1–7 (2017).
42. Samborska, B. & Kimber, M. S. A. Dodecameric CcmK2 structure suggests beta-carboxysomal shell facets have a double-layered organization. *Structure 20*, 1353–1362 (2012).
43. Sutter, M., Greber, B., Aussignargues, C. & Kerfeld, C. A. Assembly principles and structure of a 6.5-MDa bacterial microcompartment shell. *Science 356*, 1293–1297 (2017). *This study shows that the structure of a complete BMC shell reveals specific interactions among shell proteins.*
44. Faulkner, M. et al. Direct characterization of the native structure and mechanics of cyanobacterial carboxysomes. *Nanoscale 9*, 10662–10673 (2017). *This study describes the purification and structural characterization of β-carboxysomes.*
45. Kinney, J. N., Salmeen, A., Cai, F. & Kerfeld, C. A. Elucidating essential role of conserved carboxysomal protein CcmN reveals common feature of bacterial microcompartment assembly. *J. Biol. Chem. 287*, 17729–17736 (2012). *This study identifies a peptide extension for targeting in β-carboxysomes and predicts counterparts in diverse BMCs.*
46. Cameron, J. C., Wilson, S. C., Bernstein, S. L. & Kerfeld, C. A. Biogenesis of a bacterial organelle: the carboxysome assembly pathway. *Cell 155*, 1131–1140 (2013). *This study describes the assembly pathway of β-carboxysomes using knockout mutants and fluorescently labelled reporters.*

47. Ludwig, M., Sultemeyer, D. & Price, G. D. Isolation of ccmKLMN genes from the marine cyanobacterium, *Synechococcus* sp PCC7002 (Cyanobacteria), and evidence that CcmM is essential for carboxysome assembly. *J. Phycol. 36*, 1109–1118 (2000).
48. Emlyn-Jones, D., Woodger, F. J., Andrews, T. J., Price, G. D. & Whitney, S. M. A. *Synechococcus* PCC7942 ΔccmM (Cyanophyceae) mutant pseudoreverts to air growth without regaining carboxysomes. *J. Phycol. 42*, 769–777 (2006).
49. Woodger, F. J., Badger, M. R. & Price, G. D. Sensing of inorganic carbon limitation in *Synechococcus* PCC7942 is correlated with the size of the internal inorganic carbon pool and involves oxygen. *Plant Physiol. 139*, 1959–1969 (2005).
50. Berry, S., Fischer, J. H., Kruip, J., Hauser, M. & Wildner, G. F. Monitoring cytosolic pH of carboxysome-deficient cells of *Synechocystis* sp. PCC 6803 using fluorescence analysis. *Plant Biol. 7*, 342–347 (2005).
51. Chen, A. H., Robinson-Mosher, A., Savage, D. F., Silver, P. A. & Polka, J. K. The bacterial carbon-fixing organelle is formed by shell envelopment of preassembled cargo. *PLOS ONE 8*, e76127 (2013).
52. Havemann, G. D., Sampson, E. M. & Bobik, T. A. PduA is a shell protein of polyhedral organelles involved in coenzyme B-12-dependent degradation of 1,2-propanediol in *Salmonella enterica* serovar typhimurium LT2. *J. Bacteriol. 184*, 1253–1261 (2002).
53. Cheng, S., Sinha, S., Fan, C., Liu, Y. & Bobik, T. A. Genetic analysis of the protein shell of the microcompartments involved in coenzyme B12-dependent 1,2-propanediol degradation by *Salmonella. J. Bacteriol. 193*, 1385–1392 (2011).
54. Gonzalez-Esquer, C. R., Shubitowski, T. B. & Kerfeld, C. A. Streamlined construction of the cyanobacterial CO2-fixing organelle via protein domain fusions for use in plant synthetic biology. *Plant Cell 27*, 2637–2644 (2015). *This study describes the redesign of the enzymatic core of β-carboxysomes.*
55. Lassila, J. K., Bernstein, S. L., Kinney, J. N., Axen, S. D. & Kerfeld, C. A. Assembly of robust bacterial microcompartment shells using building blocks from an organelle of unknown function. *J. Mol. Biol. 426*, 2217–2228 (2014).
56. Fan, C. et al. Short N-terminal sequences package proteins into bacterial microcompartments. *Proc. Natl Acad. Sci. USA 107*, 7509–7514 (2010).
57. Choudhary, S., Quin, M. B., Sanders, M. A., Johnson, E. T. & Schmidt-Dannert, C. Engineered protein nano-compartments for targeted enzyme localization. *PLOS ONE 7*, e33342 (2012).
58. Cai, F., Bernstein, S. L., Wilson, S. C. & Kerfeld, C. A. Production and characterization of synthetic carboxysome shells with incorporated luminal proteins. *Plant Physiol. 170*, 1868–1877 (2016).
59. Sargent, F. et al. A synthetic system for expression of components of a bacterial microcompartment. *Microbiology 159*, 2427–2436 (2013).
60. Held, M. et al. Engineering formation of multiple recombinant Eut protein nanocompartments in *E. coli. Sci. Rep. 6*, 24359 (2016).
61. Quin, M. B., Perdue, S. A., Hsu, S. Y. & Schmidt-Dannert, C. Encapsulation of multiple cargo proteins within recombinant Eut nanocompartments. *Appl. Microbiol. Biotechnol. 100*, 9187–9200 (2016).
62. Wagner, H. J., Capitain, C. C., Richter, K., Nessling, M. & Mampel, J. Engineering bacterial microcompartments with heterologous enzyme cargos. *Engineer. Life Sci. 17*, 36–46 (2017).
63. Jakobson, C. M., Slininger Lee, M. F. & Tullman-Ercek, D. De novo design of signal sequences to localize cargo to the 1,2-propanediol utilization microcompartment. *Protein Sci. 26*, 1086–1092 (2017).
64. Jakobson, C. M., Kim, E. Y., Slininger, M. F., Chien, A. & Tullman-Ercek, D. Localization of proteins to the 1,2-propanediol utilization microcompartment by non-native signal sequences is mediated by a common hydrophobic motif. *J. Biol. Chem. 290*, 24519–24533 (2015).
65. Ortega, C. *et al. Mycobacterium tuberculosis* Ser/Thr protein kinase B mediates an oxygen-dependent replication switch. *PLoS Biol. 12*, e1001746 (2014).

66. Boutte, C. C. *et al.* A cytoplasmic peptidoglycan amidase homologue controls mycobacterial cell wall synthesis. *eLife 5*, e14590 (2016).
67. Holtje, J. V. Growth of the stress-bearing and shape-maintaining murein sacculus of *Escherichia coli. Microbiol. Mol. Biol. Rev. 62*, 181–203 (1998).
68. Schleifer, K. H. & Kandler, O. Peptidoglycan types of bacterial cell walls and their taxonomic implications. *Bacteriol. Rev. 36*, 407–477 (1972).
69. Sadamoto, R. et al. Cell-wall engineering of living bacteria. *J. Am. Chem. Soc. 124*, 9018–9019 (2002).
70. Liechti, G. W. et al. A new metabolic cell-wall labelling method reveals peptidoglycan in *Chlamydia trachomatis*. *Nature 506*, 507–510 (2014).
71. Kuru, E. et al. *In Situ* probing of newly synthesized peptidoglycan in live bacteria with fluorescent D-amino acids. *Angew. Chem. Int. Ed. Engl. 51*, 12519–12523 (2012).
72. Lebar, M. D. et al. Reconstitution of peptidoglycan cross-linking leads to improved fluorescent probes of cell wall synthesis. *J. Am. Chem. Soc. 136*, 10874–10877 (2014).
73. Gale, R. T. & Brown, E. D. New chemical tools to probe cell wall biosynthesis in bacteria. *Curr. Opin. Microbiol. 27*, 69–77 (2015).
74. Siegrist, M. S. et al. (D)-Amino acid chemical reporters reveal peptidoglycan dynamics of an intracellular pathogen. *ACS Chem. Biol. 8*, 500–505 (2013).
75. Shieh, P., Siegrist, M. S., Cullen, A. J. & Bertozzi, C. R. Imaging bacterial peptidoglycan with near-infrared fluorogenic azide probes. *Proc. Natl Acad. Sci. USA 111*, 5456–5461 (2014).
76. Tiyanont, K. et al. Imaging peptidoglycan biosynthesis in *Bacillus subtilis* with fluorescent antibiotics. *Proc. Natl Acad. Sci. USA 103*, 11033–11038 (2006).
77. Kocaoglu, O. et al. Selective penicillin-binding protein imaging probes reveal substructure in bacterial cell division. *ACS Chem. Biol. 7*, 1746–1753 (2012).
78. Daniel, R. A. & Errington, J. Control of cell morphogenesis in bacteria: two distinct ways to make a rod-shaped cell. *Cell 113*, 767–776 (2003).
79. Nelson, J. W. et al. A biosynthetic strategy for re-engineering the *Staphylococcus aureus* cell wall with non-native small molecules. *ACS Chem. Biol. 5*, 1147–1155 (2010).
80. Garner, E. C. et al. Coupled, circumferential motions of the cell wall synthesis machinery and MreB filaments in B. subtilis. *Science 333*, 222–225 (2011).
81. Schindler, M. K. et al. *Yersinia enterocolitica* YadA mediates complement evasion by recruitment and inactivation of C3 products. *J. Immunol. 189*, 4900–4908 (2012).
82. Koenigs, A. et al. CipA of *Acinetobacter baumannii* is a novel plasminogen binding and complement inhibitory protein. *J. Infect. Dis. 213*, 1388–1399 (2016).
83. Maruvada, R., Prasadarao, N. V. & Rubens, C. E. Acquisition of factor H by a novel surface protein on group B *Streptococcus* promotes complement degradation. *FASEB J. 23*, 3967–3977 (2009).
84. Clay, C. D., Soni, S., Gunn, J. S. & Schlesinger, L. S. Evasion of complement-mediated lysis and complement C3 deposition are regulated by *Francisella tularensis* lipopolysaccharide O antigen. *J. Immunol. 181*, 5568–5578 (2008).
85. Tan, L. A., Yang, A. C., Kishore, U. & Sim, R. B. Interactions of complement proteins C1q and factor H with lipid A and *Escherichia coli*: further evidence that factor H regulates the classical complement pathway. *Protein Cell. 2*, 320–332 (2011).
86. Smith, A. E. et al. PagP activation in the outer membrane triggers R3 core oligosaccharide truncation in the cytoplasm of *Escherichia coli* O157: H7. *J. Biol. Chem. 283*, 4332–4343 (2008).
87. Ho, D. K., Jarva, H. & Meri, S. Human complement factor H binds to outer membrane protein Rck of *Salmonella*. *J. Immunol. 185*, 1763–1769 (2010).
88. MacLennan, C. A. et al. Dysregulated humoral immunity to nontyphoidal Salmonella in HIV-infected African adults. *Science 328*, 508–512 (2010).
89. Diao, J. et al. Peptidoglycan association of murein lipoprotein is required for KpsD-dependent group 2 capsular polysaccharide expression and serum resistance in a uropathogenic *Escherichia coli* isolate. *MBio. 8*, e00603-17 (2017).

90. Coggon, C. F. et al. A novel method of serum resistance by *Escherichia coli* that causes urosepsis. *MBio 9*, e00920–18 (2018).
91. Wells, T. J. et al. Increased severity of respiratory infections associated with elevated anti-LPS IgG2 which inhibits serum bactericidal killing. *J. Exp. Med. 211*, 1893–1904 (2014).
92. Crémet, L. et al. Innate immune evasion of *Escherichia coli* clinical strains from orthopedic implant infections. *Eur. J. Clin. Microbiol Infect. Dis. 35*, 993–999 (2016).
93. Huja, S. et al. Fur is the master regulator of the extraintestinal pathogenic *Escherichia coli* response to serum. *Mbio 5*, e01460–14 (2014).
94. Hu, Y. H. & Sun, L. The global regulatory effect of *Edwardsiella tarda* Fur on iron acquisition, stress resistance, and host infection: a proteomics-based interpretation. *J. Proteom. 140*, 100–110 (2016).
95. McKellar, Stuart W.; Ivanova, Ivayla; Arede, Pedro; Zapf, Rachel L.; Mercier, Noémie; Chu, Liang-Cui; Mediati, Daniel G.; Pickering, Amy C.; Briaud, Paul; Foster, Robert G.; Kudla, Grzegorz; Fitzgerald, J. Ross; Caldelari, Isabelle; Carroll, Ronan K.; Tree, Jai J.; Granneman, Sander. RNase III CLASH in MRSA uncovers sRNA regulatory networks coupling metabolism to toxin expression. Nature Communications (2022). https://doi.org/10.1038/s41467-022-31173-y
96. Michonneau, David; Latis, Eleonora; Curis, Emmanuel; Dubouchet, Laetitia; Ramamoorthy, Sivapriya; Ingram, Brian; de Latour, Régis Peffault; Robin, Marie; de Fontbrune, Flore Sicre; Chevret, Sylvie; Rogge, Lars; Socié, Gérard. Metabolomics analysis of human acute graft-versus-host disease reveals changes in host and microbiota-derived metabolites. Nature Communications (2019). https://doi.org/10.1038/s41467-019-13498-3
97. Chiang, Nan; Fredman, Gabrielle; Bäckhed, Fredrik; Oh, Sungwhan F.; Vickery, Thad; Schmidt, Birgitta A.; Serhan, Charles N. Infection regulates pro-resolving mediators that lower antibiotic requirements. Nature (2012). https://doi.org/10.1038/nature11042
98. Vojinovic, Dina; Radjabzadeh, Djawad; Kurilshikov, Alexander; Amin, Najaf; Wijmenga, Cisca; Franke, Lude; Ikram, M. Arfan; Uitterlinden, Andre G.; Zhernakova, Alexandra; Fu, Jingyuan; Kraaij, Robert; van Duijn, Cornelia M. Relationship between gut microbiota and circulating metabolites in population-based cohorts. Nature Communications (2019). https://doi.org/10.1038/s41467-019-13721-1
99. Kerfeld, Cheryl A.; Aussignargues, Clement; Zarzycki, Jan; Cai, Fei; Sutter, Markus. Bacterial microcompartments. Nature Reviews Microbiology (2018). https://doi.org/10.1038/nrmicro.2018.10
100. Eoh, Hyungjin; Wang, Zhe; Layre, Emilie; Rath, Poonam; Morris, Roxanne; Branch Moody, D.; Rhee, Kyu Y. Metabolic anticipation in Mycobacterium tuberculosis. Nature Microbiology (2017). https://doi.org/10.1038/nmicrobiol.2017.84
101. Liang, Hai; DeMeester, Kristen E.; Hou, Ching-Wen; Parent, Michelle A.; Caplan, Jeffrey L.; Grimes, Catherine L. Metabolic labelling of the carbohydrate core in bacterial peptidoglycan and its applications. Nature Communications (2017). https://doi.org/10.1038/ncomms15015
102. Cheng, Zhi-xue; Guo, Chang; Chen, Zhuang-gui; Yang, Tian-ci; Zhang, Jian-ying; Wang, Jie; Zhu, Jia-xin; Li, Dan; Zhang, Tian-tuo; Li, Hui; Peng, Bo; Peng, Xuan-xian. Glycine, serine and threonine metabolism confounds efficacy of complement-mediated killing. Nature Communications (2019). https://doi.org/10.1038/s41467-019-11129-5

Chapter 2
Role of Various Metabolic Pathways in Bacterial Survival and Virulence

Vijay Soni and Aditya Upadhyay

2.1 Introduction by the Editor

Bacterial survival and virulence are governed by a network of metabolic pathways that extend far beyond energy production and biosynthesis. These pathways serve as dynamic interfaces between environmental sensing, stress adaptation, and pathogenic behavior. This chapter explores how diverse metabolic processes—ranging from central carbon metabolism to coenzyme stability, polyamine regulation, and nitrogen assimilation—contribute to bacterial fitness and virulence across a spectrum of pathogens.

The chapter opens with a study on *Pseudomonas aeruginosa*, revealing that core metabolic genes exert growth-independent control over virulence. This challenges the traditional view that virulence is primarily regulated by dedicated pathogenicity factors, instead highlighting the role of metabolic gene expression in shaping infection outcomes. Similarly, the longevity of coenzymes in *Escherichia coli* underscores the efficiency of microbial systems in maintaining metabolic homeostasis, allowing minimal de novo synthesis and ensuring continuity across generations.

Polyamines emerge as key modulators of metabolism and virulence in *Streptococcus pneumoniae*, influencing gene expression and stress responses. Their role in pneumococcal pathogenesis suggests that metabolic intermediates can act as regulatory signals, fine-tuning bacterial behavior in host environments. This theme is echoed in studies on sulfate-reducing bacteria (SRB), where metabolites—not cells—play dominant roles in modulating oxygen reduction reactions, with implications for microbial fuel cell technologies and environmental microbiology.

V. Soni (✉)
Department of Medicine, Weill Cornell Medicine, New York, NY, USA
e-mail: vis2032@med.cornell.edu

A. Upadhyay
Hospital for Special Surgery, New York, NY, USA

V. Soni, A. Upadhyay (eds.), *Metabolism of Bacterial Infection*,
https://doi.org/10.1007/978-3-031-46895-7_2

Iron metabolism, a critical determinant of bacterial survival, is explored through the ferric uptake regulator (Fur) in *Riemerella anatipestifer*. Fur's dual role in iron homeostasis and oxidative stress resistance demonstrates how metabolic regulation interconnects with immune evasion and virulence. In *Haemophilus influenzae*, sialic acid metabolism and its transcriptional regulation affect lipopolysaccharide sialylation, a key factor in immune evasion and serum resistance.

The global transcriptional regulator CodY in Gram-positive bacteria exemplifies how nutrient sensing integrates with virulence control. CodY's graded response to nutrient availability allows pathogens like *Staphylococcus aureus* to transition between commensal and invasive states, adapting their pathogenic strategy to environmental cues. This regulatory flexibility is further confirmed in *Brucella melitensis,* where OtpR modulates gene expression under acidic stress, coordinating metabolic and virulence pathways to ensure intracellular survival.

Carbon metabolism in intracellular pathogens such as *Listeria monocytogenes*, *Shigella flexneri*, and *Mycobacterium tuberculosis* reveals pathogen-specific adaptations to host environments. These bacteria exploit host-derived carbon sources—glycerol, glucose, fatty acids—to fuel replication and modulate virulence gene expression. In EHEC, fumarate emerges as a central metabolic regulator, linking the TCA cycle to virulence through transcriptional control.

Stress adaptation mechanisms are further explored in the context of cell structure and group behavior. Bacteria employ structural defenses, metabolic reprogramming, and quorum sensing to survive hostile conditions. Mucin glycans, as host-derived regulators, attenuate *P. aeruginosa* virulence by disrupting biofilms and suppressing quorum sensing, offering insights into non-antibiotic therapeutic strategies.

Finally, the stringent response mediated by alarmones such as ppGpp and pGpp illustrates how bacteria reprogram their physiology under nutrient starvation. These signaling molecules orchestrate global changes in gene expression, metabolism, and stress resistance, highlighting their importance in chronic infections and survival under hostile environments.

Together, these studies reveal that bacterial metabolism is not a passive background process but a central determinant of pathogenic potential. By dissecting the roles of various metabolic pathways, this chapter provides a comprehensive view of how bacteria integrate environmental signals, metabolic states, and regulatory networks to survive and thrive in diverse host niches.

2.2 Machine Generated Summaries

Disclaimer: The summaries in this chapter were generated from Springer Nature publications using abstractive AI auto-summarization: An abstraction-based summarizer creates new text based on deep learning. New phrases are created to summarize the content. As the constituted sentences are machine selected, they may not fully reflect the body of the work, so we strongly advise that the original content is read and cited. The auto-generated summaries were curated by the editor to meet Springer Nature publication standards.

Pseudomonas aeruginosa *Core Metabolism Exerts a Widespread Growth-Independent Control on Virulence [159]*

This is a machine-generated summary of:

Panayidou, Stavria; Georgiades, Kaliopi; Christofi, Theodoulakis; Tamana, Stella; Promponas, Vasilis J.; Apidianakis, Yiorgos Pseudomonas aeruginosa Core Metabolism Exerts a Widespread Growth-Independent Control on Virulence [159]

Published in: Scientific Reports (2020)

Link to original: https://doi.org/10.1038/s41598-020-66194-4

If you want to cite the papers, please refer to the original.

For technical reasons we could not place the page where the original quote is coming from.

Research Highlights

Pseudomonas species are versatile bacteria capable of inhabiting diverse environments, with some strains, such as *Pseudomonas aeruginosa*, being notable opportunistic human pathogens. These bacteria can cause acute infections in immunocompromised individuals and persistent respiratory infections in cystic fibrosis patients. Despite sharing a significant portion of their gene content, *P. aeruginosa* strains exhibit vast differences in virulence that cannot be solely explained by the presence of known virulence factors. Now, a new study by Stavria Panayidou and colleagues explores the role of gene expression, particularly of virulence factors and core metabolism genes, in explaining these differences in pathogenicity among *Pseudomonas* strains.

The study involved large-scale comparisons of fully sequenced *Pseudomonas* genomes at the protein sequence level to investigate the correlation between gene expression and pathogenicity. The researchers utilized various infection models, including *Drosophila melanogaster* and mouse lung infection assays, to assess the virulence of different *Pseudomonas* strains. They employed RNA sequencing to analyze gene expression profiles and identified differentially expressed genes between highly and lowly virulent strains. The study found that while the presence or absence of specific virulence factors did not correlate with pathogenicity, the expression levels of certain virulence genes did. Additionally, core metabolism genes were found to affect virulence at the pathway level rather than individually.

The findings suggest that differences in *Pseudomonas aeruginosa* virulence are influenced by the expression of specific virulence factors and core metabolic pathways rather than just their presence or absence. This highlights the complexity of bacterial pathogenicity and suggests that future studies could focus on understanding the regulatory mechanisms controlling gene expression in these pathways. Such insights could lead to novel therapeutic strategies targeting metabolic pathways to mitigate *P. aeruginosa* infections without affecting bacterial growth, potentially reducing resistance development.

Longevity of Major Coenzymes Allows Minimal De Novo Synthesis in Microorganisms [160]

This is a machine-generated summary of:

Hartl, Johannes; Kiefer, Patrick; Meyer, Fabian; Vorholt, Julia A. Longevity of Major Coenzymes Allows Minimal De Novo Synthesis in Microorganisms [160]

Published in: Nature Microbiology (2017)

Link to original: https://doi.org/10.1038/nmicrobiol.2017.73

If you want to cite the papers, please refer to the original.

For technical reasons we could not place the page where the original quote is coming from.

Research Highlights

Enzymes rely heavily on organic cofactors, or coenzymes, to expand the biochemical capabilities of cellular metabolism beyond what proteins alone can achieve. These coenzymes, such as nicotinamide adenine dinucleotide (NAD), are crucial for the function of oxidoreductases and other enzyme classes. Despite their importance, the in vivo stability and turnover rates of these coenzymes at the cellular level remain poorly understood. Now, a new study by Johannes Hartl and colleagues addresses this knowledge gap by investigating the turnover of organic coenzymes in *Escherichia coli* using metabolic tracer analysis.

The researchers employed metabolic tracer analysis with carbon-13 labeling to track the incorporation and depletion of isotopes in coenzymes over time. They used liquid chromatography–high-resolution mass spectrometry (LC-HRMS) to measure isotopologue distributions, which allowed them to determine the pool turnover rates of various coenzymes. By switching between carbon-12 and carbon-13 labeled substrates, they could assess how quickly coenzymes were synthesized anew or recycled within the cell. The study revealed that most coenzymes are produced primarily to compensate for dilution due to cell division rather than active turnover, indicating a high degree of chemical stability and homeostasis in coenzyme metabolism.

The findings suggest that *E. coli* maintains coenzyme pools with minimal de novo synthesis, relying instead on effective salvage pathways and repair mechanisms to preserve these essential molecules. This stability allows for efficient propagation of coenzymes across generations without significant loss or damage. Future study could explore how these mechanisms are conserved across different organisms and under varying environmental conditions. Additionally, understanding how cells regulate coenzyme homeostasis could provide insights into metabolic diseases where these processes are disrupted. The study highlights the remarkable efficiency of cellular systems in managing coenzyme chemistry and ensuring metabolic continuity.

The Expansive Effects of Polyamines on the Metabolism and Virulence of **Streptococcus pneumoniae** *[161]*

This is a machine-generated summary of:

Nanduri, Bindu; Swiatlo, Edwin The Expansive Effects of Polyamines on the Metabolism and Virulence of Streptococcus pneumoniae [161]

Published in: Pneumonia (2021)

Link to original: https://doi.org/10.1186/s41479-021-00082-x

If you want to cite the papers, please refer to the original.

For technical reasons we could not place the page where the original quote is coming from.

Research Highlights

Lower respiratory tract infections, particularly pneumonia, remain a significant global health challenge, causing high morbidity and mortality rates, especially in children under 5 and across all age groups. *Streptococcus pneumoniae* is the most common bacterial cause of pneumonia, but the effectiveness of current vaccines is limited by the diversity of pneumococcal serotypes and the emergence of antibiotic-resistant strains. Now, a new study by Bindu Nanduri and colleagues explores the role of polyamines in pneumococcal physiology and pathogenesis, identifying potential new targets for intervention.

The study investigates the metabolic pathways and regulatory mechanisms of polyamines in *Streptococcus pneumoniae*. Polyamines are small molecules involved in various cellular functions, including interactions with nucleic acids that affect transcription and translation. The researchers focused on genes involved in polyamine biosynthesis and transport within pneumococcal genomes. They identified a single ATP-binding cassette transporter (PotABCD) responsible for polyamine uptake and explored its role in pneumococcal growth and virulence. Using gene deletion mutants, they demonstrated that while polyamine synthesis is not essential for growth in vitro, it significantly impacts virulence in vivo. The study also examined how polyamine metabolism affects pneumococcal stress responses and interactions with host immune systems.

The findings suggest that polyamine metabolism plays a crucial role in pneumococcal pathogenesis by regulating gene expression and central metabolism. This regulation impacts the bacterium's ability to adapt to different host environments, influencing its virulence. Future study could focus on further elucidating the mechanisms by which polyamines affect pneumococcal physiology and exploring their potential as targets for new therapeutic strategies. The research highlights the importance of understanding metabolic pathways in bacterial pathogens to develop more effective treatments and preventive measures against diseases like pneumonia.

Oxygen Reduction Reaction Affected by Sulfate-Reducing Bacteria: Different Roles of Bacterial Cells and Metabolites [162]

This is a machine-generated summary of:

Wu, Jiajia; Liu, Huaiqun; Wang, Peng; Zhang, Dun; Sun, Yan; Li, Ee Oxygen Reduction Reaction Affected by Sulfate-Reducing Bacteria: Different Roles of Bacterial Cells and Metabolites [162]

Published in: Indian Journal of Microbiology (2017)
Link to original: https://doi.org/10.1007/s12088-017-0667-z

If you want to cite the papers, please refer to the original.

For technical reasons we could not place the page where the original quote is coming from.

Research Highlights

Microbial biocathodes for the oxygen reduction reaction (ORR) are commonly used in microbial fuel cells due to their cost efficiency, sustainability, and resistance to poisoning compared to conventional chemical catalysts. While both mixed microbial communities and pure cultures have been reported to catalyze ORR, the complex nature of mixed communities complicates mechanism investigations, prompting research into pure cultures. Although the catalytic activity of ORR has been identified in several bacterial strains, such as *Roseobacter* sp. and *Yersinia enterocolitica*, it remains unclear how anaerobes like sulfate-reducing bacteria (SRB) influence ORR. Now, a new study by Jiajia Wu and colleagues explores the impact of SRB and their metabolites on ORR.

The study involved isolating an SRB strain from marine sludge in Huiquan Bay, China, using a modified Postgate's medium for cultivation. The researchers performed phylogenetic analysis to confirm the strain's identity as part of the Desulfovibrio group. To assess the impact on ORR, cyclic voltammetry experiments were conducted using a three-electrode system with glassy carbon electrodes in nitrogen- and oxygen-saturated NaCl electrolytes containing varying concentrations of SRB culture solution, bacterial cells, and metabolites. The study found that SRB culture solutions significantly affected ORR processes by decreasing current at specific cathodic peaks. Further investigation revealed that metabolites had a more substantial impact than bacterial cells, with sulfide affecting the reduction of hydrogen peroxide (H_2O_2) due to pH changes and chemical reactions, while extracellular polymeric substances (EPS) influenced both the reduction of oxygen to superoxide anion and H_2O_2.

The findings demonstrate that SRB culture solutions significantly impact ORR, with metabolites playing a more crucial role than bacterial cells. Sulfide was found to heavily affect the third reduction step of ORR due to its hydrolysis and reaction with H_2O_2, while EPS hindered oxygen reduction but catalyzed H_2O_2 reduction. Future studies could explore other SRB metabolites' roles in ORR and their cooperative actions for a deeper understanding of interactions between oxygen and SRB. This research provides insights into microbial biocathodes' potential applications in sustainable energy technologies.

Functional Characterization of Fur in Iron Metabolism, Oxidative Stress Resistance and Virulence of Riemerella anatipestifer *[163]*

This is a machine-generated summary of:

Huang, Mi; Liu, Mafeng; Liu, Jiajun; Zhu, Dekang; Tang, Qianying; Jia, Renyong; Chen, Shun; Zhao, Xinxin; Yang, Qiao; Wu, Ying; Zhang, Shaqiu; Huang, Juan; Ou, Xumin; Mao, Sai; Gao, Qun; Sun, Di; Wang, Mingshu; Cheng, Anchun Functional Characterization of Fur in Iron Metabolism, Oxidative Stress Resistance and Virulence of *Riemerella anatipestifer* [163]

Published in: Veterinary Research (2021)

Link to original: https://doi.org/10.1186/s13567-021-00919-9

If you want to cite the papers, please refer to the original.

For technical reasons we could not place the page where the original quote is coming from.

Research Highlights

Iron is a crucial element for bacterial survival, playing a vital role in various metabolic processes. However, excess iron can lead to the production of toxic reactive oxygen species (ROS) through the Fenton reaction, causing cellular damage. To manage iron levels, bacteria have developed regulatory systems such as the ferric uptake regulator (Fur), which controls iron homeostasis by modulating gene expression based on intracellular iron concentrations. *Riemerella anatipestifer*, a Gram-negative bacterium affecting avian species, presents challenges due to its multiple serotypes and antibiotic resistance. Although previous studies have highlighted the importance of iron for *R. anatipestifer*'s survival and pathogenesis, the specific role of Fur in this bacterium remains unclear. Now, a new study by Mi Huang and colleagues investigates the function of Fur in maintaining iron homeostasis, oxidative stress resistance, and pathogenesis in *R. anatipestifer* CH-1.

The study employed various experimental approaches to elucidate Fur's role in *R. anatipestifer* CH-1. Researchers constructed mutant strains lacking the Fur gene and assessed their growth under different iron conditions using media supplemented with an iron chelator. They performed sensitivity assays to evaluate the mutants' response to oxidative stress induced by hydrogen peroxide (H_2O_2) and streptonigrin, an antibiotic whose efficacy is enhanced by iron presence. Additionally, they measured intracellular ROS levels using fluorescence-based detection methods and conducted electrophoretic mobility shift assays (EMSAs) to determine Fur's binding affinity to promoter regions of putative iron uptake genes. The study also explored Fur's contribution to virulence through duckling infection models, assessing bacterial colonization and survival rates in host tissues.

The findings reveal that Fur plays a critical role in regulating iron uptake genes by binding to their promoters, thereby preventing excessive intracellular iron accumulation and subsequent oxidative stress. The Fur mutant exhibited increased sensitivity to H_2O_2 and streptonigrin due to elevated ROS levels, highlighting Fur's importance in oxidative stress resistance. Furthermore, the absence of Fur resulted in reduced virulence and colonization ability in ducklings, suggesting its involvement in pathogenesis. Future studies could explore whether Fur regulates oxidative stress response enzymes or virulence factors directly. These insights into Fur's multifaceted role provide a foundation for developing potential attenuated vaccines against *R. anatipestifer* infections.

Sialic Acid-Mediated Transcriptional Modulation of a Highly Conserved Sialometabolism Gene Cluster in **Haemophilus influenzae** *and Its Effect on Virulence [164]*

This is a machine-generated summary of:

Jenkins, Gaynor A; Figueira, Marisol; Kumar, Gaurav A; Sweetman, Wendy A; Makepeace, Katherine; Pelton, Stephen I; Moxon, Richard; Hood, Derek W Sialic Acid-Mediated Transcriptional Modulation of a Highly Conserved Sialometabolism Gene Cluster in *Haemophilus influenzae* and Its Effect on Virulence [164]

Published in: BMC Microbiology (2010)

Link to original: https://doi.org/10.1186/1471-2180-10-48

If you want to cite the papers, please refer to the original.

For technical reasons we could not place the page where the original quote is coming from.

Research Highlights

Sialic acid is a crucial sugar found on glycoconjugates in many animals, playing significant roles in cell–cell adhesion, complement activation prevention, and tissue function regulation. In *Haemophilus influenzae*, a Gram-negative bacterium residing in the human respiratory tract, sialylation of its major surface glycolipid, lipopolysaccharide (LPS), is linked to its virulence and ability to evade the host immune system. Despite its importance, the regulatory mechanisms governing sialic acid utilization and LPS sialylation in *H. influenzae* remain poorly understood. Now, a new study by Gaynor A. Jenkins and colleagues investigates the transcriptional regulation of sialometabolism genes in *H. influenzae*, revealing complex regulatory processes that coordinate sialic acid uptake and catabolism.

The study employed reverse transcriptase PCR to examine the transcription of sialometabolism genes in both wild-type and mutant strains of *H. influenzae* grown with or without added sialic acid. The researchers constructed mutant strains by disrupting key genes involved in sialic acid uptake and catabolism, such as siaP and siaR, using PCR cloning techniques. These mutants were analyzed for changes in LPS structure through electrophoresis and assessed for serum resistance using bactericidal assays. Additionally, gene expression was quantified using real-time quantitative PCR (q-PCR) to understand the regulatory roles of SiaR and CRP proteins. The study also utilized a chinchilla model of otitis media to evaluate the virulence of these mutants compared to wild-type strains.

The findings highlight that sialometabolism gene expression is intricately regulated by both SiaR and CRP proteins, with SiaR acting as a repressor and CRP primarily enhancing uptake gene expression. Mutations in these regulatory genes did not significantly affect virulence in the chinchilla model, suggesting that LPS sialylation remains intact despite altered gene expression. Future studies could explore infection models initiated through nasopharyngeal inoculation to better understand the evolutionary pressures on LPS sialylation related to *H. influenzae*'s fitness for carriage and transmission. Understanding these mechanisms may offer new targets for interventions against diseases caused by this bacterium.

CodY, a Master Integrator of Metabolism and Virulence in Gram-Positive Bacteria [165]

This is a machine-generated summary of:

Brinsmade, Shaun R. CodY, a Master Integrator of Metabolism and Virulence in Gram-Positive Bacteria [165]

Published in: Current Genetics (2016)

Link to original: https://doi.org/10.1007/s00294-016-0656-5

If you want to cite the papers, please refer to the original.

For technical reasons we could not place the page where the original quote is coming from.

Research Highlights

Prokaryotes exhibit remarkable adaptability to diverse environments, largely due to protein and RNA-based factors that sense environmental changes and adjust gene expression accordingly. Among these factors, the global transcriptional regulator CodY plays a crucial role in Gram-positive bacteria by modulating gene expression in response to nutrient availability. Despite extensive research on CodY's function, the precise mechanisms by which it integrates metabolic signals with virulence regulation remain unclear. Now, a new study by Shaun R. Brinsmade and colleagues explores how CodY regulates gene expression in *Staphylococcus aureus*, shedding light on its role in balancing commensalism and pathogenicity.

The study utilized a collection of *S. aureus* strains with varying levels of CodY activity to investigate how this regulator influences gene expression under different nutrient conditions. By examining mutant strains with altered isoleucine, leucine, valine (ILV)-binding pockets, the researchers mimicked scenarios of nutrient limitation, revealing that CodY prioritizes gene expression based on nutrient availability. The team discovered that as nutrients become scarce, CodY activity decreases, leading to the derepression of virulence genes that facilitate tissue invasion and immune evasion. This graded response allows *S. aureus* to adapt its pathogenic strategy according to environmental cues, highlighting CodY's role as a master integrator of metabolism and virulence.

The findings underscore CodY's significance in controlling virulence in Gram-positive pathogens like *S. aureus*. Future studies could focus on unraveling the molecular mechanisms by which CodY interacts with other key regulators such as Agr, Rot, and Sae to integrate host-derived signals into the virulence regulatory network. Additionally, understanding how CodY activity is modulated during infection could provide insights into the commensal-to-pathogen switch in vivo. These insights may pave the way for novel therapeutic strategies aimed at maintaining CodY activity to prevent *S. aureus* from transitioning to a pathogenic state, offering potential solutions in an era of increasing antibiotic resistance.

RNA-Seq Reveals the Critical Role of OtpR in Regulating **Brucella melitensis** *Metabolism and Virulence Under Acidic Stress [166]*

This is a machine-generated summary of:

Liu, Wenxiao; Dong, Hao; Li, Jing; Ou, Qixing; Lv, Yujin; Wang, Xiaolei; Xiang, Zuoshuang; He, Yongqun; Wu, Qingmin RNA-Seq Reveals the Critical Role of

OtpR in Regulating *Brucella melitensis* Metabolism and Virulence Under Acidic Stress [166]

Published in: Scientific Reports (2015)
Link to original: https://doi.org/10.1038/srep10864

If you want to cite the papers, please refer to the original.

For technical reasons we could not place the page where the original quote is coming from.

Research Highlights

Brucella spp. are facultative intracellular bacteria whose virulence is largely dependent on their ability to survive within phagocytes, which present harsh environments such as low pH and limited nutrients. The transcription regulator OtpR has been identified as a key player in regulating stress responses, cell growth, and morphology in *Brucella*. However, the full extent of genes and networks regulated by OtpR, particularly under acidic conditions within macrophages, remains unclear. Now, a new study by Wenxiao Liu and colleagues employs RNA sequencing to elucidate the gene expression profiles regulated by OtpR in *Brucella melitensis* under acid stress.

The study utilized RNA-seq technology to compare the transcriptomes of the wild-type *B. melitensis* strain M and its otpR mutant counterpart MΔotpR under acidic conditions. The researchers identified differentially expressed genes by aligning sequence reads to the reference genome and quantifying expression levels using the RPKM algorithm. They found that OtpR regulates a wide array of genes involved in critical cellular processes such as carbohydrate metabolism, nitrogen metabolism, iron metabolism, cell division, and virulence factor expression. Notably, genes associated with the tricarboxylic acid cycle and Type IV secretion system were significantly down-regulated in the otpR mutant. The study also confirmed these findings through RT-qPCR assays on selected genes.

This research advances our understanding of *Brucella* pathogenesis by highlighting OtpR's role in regulating metabolic pathways and virulence factors under acidic stress. The findings suggest that OtpR may indirectly regulate virulence genes through interactions with other transcriptional regulators like VjbR and BlxR. Future studies could focus on identifying the sensor component of the OtpR regulatory system and mapping its interaction networks more comprehensively. Such insights could inform strategies for developing therapeutic interventions against *Brucella* infections by targeting these regulatory pathways.

Carbon Metabolism of Intracellular Bacterial Pathogens and Possible Links to Virulence [167]

This is a machine-generated summary of:

Eisenreich, Wolfgang; Dandekar, Thomas; Heesemann, Jürgen; Goebel, Werner Carbon Metabolism of Intracellular Bacterial Pathogens and Possible Links to Virulence [167]

Published in: Nature Reviews Microbiology (2010)

Link to original: https://doi.org/10.1038/nrmicro2351

If you want to cite the papers, please refer to the original.

For technical reasons we could not place the page where the original quote is coming from.

Research Highlights

Intracellular bacterial pathogens, such as *Listeria monocytogenes, Shigella flexneri, Salmonella enterica* serovar Typhimurium, and *Mycobacterium tuberculosis*, have evolved complex mechanisms to adapt to the host environment, facilitating their survival and replication. These adaptations include specific adherence to host cells, modification of vacuolar compartments, and modulation of immune responses. While the acquisition of virulence factors has been extensively studied, the metabolic adaptations of these pathogens within host cells remain less understood. Now, a new study by Wolfgang Eisenreich and colleagues provides insights into the intracellular carbon metabolism of these pathogens and its potential links to virulence.

The study employs various advanced techniques such as differential gene expression profiling (DGEP) and C-isotopologue-profiling analysis (C-IPA) to investigate the metabolic pathways utilized by these bacteria within host cells. By comparing transcription profiles from bacteria grown in cell culture with those from infected host cells, the researchers identified key metabolic adaptations. For instance, *L. monocytogenes* primarily utilizes glycerol as a carbon source in the cytosol, while *S. flexneri* shows strain-dependent preferences for glucose or C3 substrates. In contrast, *S. Typhimurium* relies on glucose during intracellular growth in phagosomal vacuoles, whereas *M. tuberculosis* preferentially uses fatty acids or cholesterol-derived carbon units in macrophages. These findings highlight the diverse metabolic strategies employed by intracellular pathogens to exploit host cell resources.

The study advances our understanding of pathogen-specific metabolic adaptations and their potential role in virulence regulation. It suggests that intracellular bacteria exhibit flexible metabolism, allowing them to switch between different carbon sources depending on availability. This flexibility may be crucial for efficient replication and could influence the expression of virulence genes necessary for survival within host cells. Future studies using refined in vivo systems and sensitive techniques are needed to validate these findings under real infection conditions. Such research could provide deeper insights into bacterial pathogenesis and identify novel targets for antimicrobial therapies.

A Multi-Omic Analysis Reveals the Role of Fumarate in Regulating the Virulence of Enterohemorrhagic Escherichia coli *[168]*

This is a machine-generated summary of:

Kuo, Cheng-Ju; Wang, Sin-Tian; Lin, Chia-Mei; Chiu, Hao-Chieh; Huang, Cheng-Rung; Lee, Der-Yen; Chang, Geen-Dong; Chou, Ting-Chen; Chen, Jenn-Wei; Chen, Chang-Shi A Multi-Omic Analysis Reveals the Role of Fumarate in Regulating the Virulence of Enterohemorrhagic *Escherichia coli* [168]

Published in: Cell Death & Disease (2018)

Link to original: https://doi.org/10.1038/s41419-018-0423-2

If you want to cite the papers, please refer to the original.

For technical reasons we could not place the page where the original quote is coming from.

Research Highlights

Enterohemorrhagic *Escherichia coli* (EHEC) is a significant human pathogen known for causing severe gastrointestinal diseases, including diarrhea and hemolytic uremic syndrome. EHEC's virulence is influenced by various factors, such as the locus of enterocyte effacement and the F-like virulence plasmid. Despite understanding these factors, the role of bacterial central metabolism in EHEC's pathogenicity remains unclear. Now, a new study by Cheng-Ju Kuo and colleagues explores how the succinate dehydrogenase (Sdh) complex, particularly the sdhA gene, influences EHEC virulence in *Caenorhabditis elegans*.

The study utilized a transposon mutagenesis library to identify potential virulence determinants in EHEC using *C. elegans* as a model organism. By screening mutants, the researchers identified that mutations in the sdhA gene significantly reduced EHEC's pathogenicity without affecting bacterial growth. The Sdh complex, part of both the TCA cycle and aerobic respiration, was found to be crucial for full virulence expression. Complementation experiments confirmed that restoring *sdhA* function reversed the attenuated virulence phenotype. Additionally, metabolomic analysis revealed that fumarate depletion in sdhA mutants contributed to reduced pathogenicity, suggesting fumarate as a key metabolic regulator of EHEC virulence.

The findings highlight the importance of central metabolism in bacterial pathogenesis and suggest that fumarate plays a critical role in regulating EHEC's virulence through the Cra transcription factor. Future study could explore how fumarate directly or indirectly modulates Cra activity and its downstream effects on virulence gene expression. This research provides insights into potential metabolic targets for therapeutic intervention against EHEC infections, emphasizing the interplay between metabolic pathways and bacterial pathogenicity. Understanding these mechanisms could lead to novel strategies for controlling EHEC and similar pathogens by disrupting their metabolic dependencies.

The Impact of Cell Structure, Metabolism and Group Behavior for the Survival of Bacteria Under Stress Conditions [169]

This is a machine-generated summary of:

Zhang, Xinyi; Li, Zhendong; Pang, Shengmei; Jiang, Boyu; Yang, Yang; Duan, Qiangde; Zhu, Guoqiang The Impact of Cell Structure, Metabolism and Group Behavior for the Survival of Bacteria Under Stress Conditions [169]

Published in: Archives of Microbiology (2020)

Link to original: https://doi.org/10.1007/s00203-020-02050-3

If you want to cite the papers, please refer to the original.

For technical reasons we could not place the page where the original quote is coming from.

Research Highlights

Bacteria are constantly exposed to various environmental stresses, such as nutrient deprivation, temperature fluctuations, and toxic substances, which can threaten their survival. To cope with these challenges, bacteria have evolved sophisticated stress response mechanisms that involve changes in cell structures, metabolic pathways, and group behaviors. However, the precise regulatory mechanisms underlying these responses remain largely unknown. Now a new study by Xinyi Zhang and colleagues provides a comprehensive review of bacterial stress responses, focusing on the roles of cell structures, metabolic processes, and quorum sensing (QS) in enhancing bacterial resilience.

The study explores how bacteria utilize their cell envelope as a primary defense against environmental stressors and antimicrobial agents. The authors describe how Gram-negative bacteria's outer membrane (OM) and periplasmic space contribute to mechanical stability and resistance to osmotic fluctuations. They also highlight the role of ribosomes in energy conservation during nutrient deprivation by forming hibernating ribosomes that suppress protein synthesis. Additionally, the review discusses the involvement of DNA and RNA in forming stable nucleoprotein complexes that protect against oxidative and thermal stress. Metabolically, bacteria can reprogram pathways to conserve nutrients and detoxify harmful metabolites. The study also emphasizes the importance of QS in coordinating group behaviors such as biofilm formation and antibiotic resistance through signaling molecules like autoinducers.

The advances highlighted in this review underscore the complexity of bacterial stress responses and suggest potential avenues for future research. Future study could focus on unraveling the regulatory mechanisms of QS at low cell densities or exploring how sigma factors influence QS gene expression during stress conditions. Understanding these processes could inform strategies to combat bacterial infections by targeting stress response pathways or inhibiting QS to prevent biofilm formation and reduce antibiotic resistance. This comprehensive overview by Zhang et al. provides valuable insights into bacterial adaptation strategies, paving the way for novel approaches to managing microbial threats.

Mucin Glycans Attenuate the Virulence of Pseudomonas aeruginosa *in Infection [170]*

This is a machine-generated summary of:

Wheeler, Kelsey M.; Cárcamo-Oyarce, Gerardo; Turner, Bradley S.; Dellos-Nolan, Sheri; Co, Julia Y.; Lehoux, Sylvain; Cummings, Richard D.; Wozniak, Daniel J.; Ribbeck, Katharina Mucin Glycans Attenuate the Virulence of *Pseudomonas aeruginosa* in Infection [170]

Published in: Nature Microbiology (2019)

Link to original: https://doi.org/10.1038/s41564-019-0581-8

If you want to cite the papers, please refer to the original.

For technical reasons we could not place the page where the original quote is coming from.

Research Highlights

The study of mucus and its role in controlling pathogens is crucial, particularly in understanding how it influences bacterial behavior. *Pseudomonas aeruginosa*, an opportunistic pathogen, poses a significant threat to individuals with compromised immunity or disrupted mucus production. Despite its potential for causing severe infections, the mechanisms by which native mucus regulates such pathogens remain unclear. Now, a new study by Kelsey M. Wheeler and colleagues explores how native intestinal mucus affects the biofilm integrity of *P. aeruginosa*, revealing that exposure to mucus disrupts biofilms and shifts bacterial cells into a planktonic state.

Wheeler and her team conducted experiments exposing mature *P. aeruginosa* biofilms to either buffer or native intestinal mucus. While buffer alone had no impact, the presence of mucus led to biofilm dispersal and a shift in bacterial cells to a planktonic phase without altering growth rates. This dispersal was found to be dependent on an intact flagellum, suggesting an active motility-driven escape rather than mechanical disruption. Further analysis showed that intestinal mucus suppresses quorum sensing and type-secretion systems, indicating a broader regulatory effect on virulence traits. The researchers identified mucin polymers as key components responsible for these effects by testing different molecular weight fractions of mucus and confirming that purified mucins could restore biofilm dispersal and gene suppression.

The study advances our understanding of how mucins regulate bacterial behavior at both phenotypic and genetic levels. Future study could focus on identifying specific glycan structures within mucins that contribute to these regulatory effects, as well as exploring their potential therapeutic applications in treating bacterial infections. The findings suggest that mucin glycans may serve as potent host-derived regulators capable of attenuating virulence while maintaining microbial diversity, offering insights into novel strategies for infection control and microbiota stabilization.

Aspartate Aminotransferase Rv3722c Governs Aspartate-Dependent Nitrogen Metabolism in Mycobacterium tuberculosis *[171]*

This is a machine-generated summary of:

Jansen, Robert S.; Mandyoli, Lungelo; Hughes, Ryan; Wakabayashi, Shoko; Pinkham, Jessica T.; Selbach, Bruna; Guinn, Kristine M.; Rubin, Eric J.; Sacchettini, James C.; Rhee, Kyu Y. Aspartate Aminotransferase Rv3722c Governs Aspartate-Dependent Nitrogen Metabolism in *Mycobacterium tuberculosis* [171]

Published in: Nature Communications (2020)

Link to original: https://doi.org/10.1038/s41467-020-15876-8

If you want to cite the papers, please refer to the original.

For technical reasons we could not place the page where the original quote is coming from.

Research Highlights

The rapid accumulation of sequence data has outpaced the ability to annotate its information content, leaving many predicted coding sequences unannotated or misannotated. This issue is particularly problematic for microbial pathogens like **Mycobacterium tuberculosis** (Mtb), where gaps in annotation restrict access to potentially actionable features of their physiology. Despite the development of curative chemotherapies for tuberculosis (TB) decades ago, current treatments remain lengthy, complex, and toxic, contributing to treatment noncompliance and the emergence of drug resistance. Now, a new study by Robert S. Jansen and colleagues identifies a previously uncharacterized Mtb gene as the primary aspartate aminotransferase (AspAT), essential for Mtb growth and virulence.

Jansen et al. employed a combination of genetic, biochemical, and structural approaches to elucidate the function of this gene. They constructed an Mtb strain with regulated expression of the gene using a tetracycline-repressible protein degradation system. This allowed them to demonstrate that depletion of the protein resulted in impaired growth in vitro and in macrophages, as well as attenuation in a mouse model of TB infection. Through activity-based metabolomic profiling (ABMP), they identified the gene product as an aminotransferase that catalyzes the biosynthesis of aspartate (Asp) from glutamate (Glu). Structural analysis revealed that this enzyme belongs to a distinct subclass of AspATs absent in humans, suggesting potential for selective drug targeting.

The study advances our understanding of nitrogen metabolism in Mtb by highlighting the essential role of Asp biosynthesis and distribution in bacterial growth and pathogenicity. The identification of this enzyme as the main AspAT in Mtb underscores its potential as a drug target for TB treatment. Future studies could explore inhibitors targeting this enzyme class, leveraging its unique structural features absent in humans. Additionally, further research into nitrogen metabolism pathways may reveal other critical targets for therapeutic intervention against TB.

The Stringent Response and Physiological Roles of (pp)pGpp in Bacteria [172]

This is a machine-generated summary of:

Irving, Sophie E.; Choudhury, Naznin R.; Corrigan, Rebecca M. The Stringent Response and Physiological Roles of (pp)pGpp in Bacteria [172]

Published in: Nature Reviews Microbiology (2020)

Link to original: https://doi.org/10.1038/s41579-020-00470-y

If you want to cite the papers, please refer to the original.

For technical reasons we could not place the page where the original quote is coming from.

Research Highlights

Bacteria face numerous environmental stresses throughout their life cycle, necessitating rapid and efficient adaptive responses to ensure survival. Among the various stress signaling pathways, the stringent response is a critical mechanism that enables bacteria to respond to nutrient starvation by accumulating alarmones such as guanosine tetraphosphate (ppGpp). These nucleotides are synthesized by the RSH protein superfamily and play a pivotal role in regulating bacterial metabolism and physiology. Despite extensive research on ppGpp, the presence and function of a third member of this family, guanosine 3′-monophosphate 5′-diphosphate (pGpp), have only recently been confirmed in numerous species. Now, a new study by Sophie E. Irving and colleagues provides an in-depth review of recent advances in pGpp signaling, highlighting its synthesis, degradation, and regulatory roles across various bacterial species.

The study employs a comprehensive approach to elucidate the mechanisms underlying pGpp synthesis and function. Using hidden Markov model-based genome searches, the authors identified RSH enzymes responsible for synthesizing and hydrolyzing pGpp across different bacterial taxa. They classified these enzymes into subfamilies based on phylogeny and explored their structural and functional diversity. The research further delves into the interaction partners of long RSH proteins, such as SpoT from *E. coli*, which respond to specific stress signals like fatty acid starvation. Additionally, the study examines small alarmone synthetases (SASs) like RelP and RelQ, which lack regulatory sensory domains but are transcriptionally regulated under stress conditions. The authors also discuss the role of small alarmone hydrolases (SAHs) in maintaining cellular homeostasis by degrading pGpp.

The findings underscore the complexity of pGpp signaling pathways and their critical role in bacterial adaptation to stress. The study highlights that while much progress has been made in understanding pGpp's regulatory functions, many aspects remain unexplored, particularly concerning its role in chronic infections and potential functions beyond bacteria. Future studies could focus on characterizing unstudied RSH subfamilies and identifying novel enzymes involved in pGpp turnover. Additionally, exploring pGpp's presence in plant chloroplasts and green algae could reveal broader biological roles for these alarmones. The development of new tools for detecting pGpp production in vivo will be crucial for advancing our understanding of this essential signaling system's impact on bacterial pathogenicity and survival strategies.

2.3 Conclusion

Studies presented in this chapter underline the profound influence of metabolic pathways on bacterial survival, adaptation, and virulence. Far from being isolated biochemical processes, metabolism in bacteria functions as an essential regulatory hub that integrates environmental signals, stress responses, and pathogenic strategies. This chapter has highlighted how diverse metabolic systems—from central carbon metabolism to coenzyme stability, iron regulation, and quorum sensing—contribute to the dynamic interplay between microbial physiology and host interactions.

One of the key insights emerging from this chapter is the concept of growth-independent metabolic control of virulence, as demonstrated in *Pseudomonas aeruginosa*. The expression of core metabolic genes, rather than their mere presence, determines pathogenic potential, suggesting that targeting metabolic regulation could be a viable strategy for attenuating virulence without compromising bacterial viability—a promising approach for reducing selective pressure for resistance.

The longevity and stability of c in *E. coli* reveal the efficiency of microbial systems in maintaining metabolic continuity. These findings suggest that bacteria rely heavily on salvage and repair mechanisms rather than constant synthesis, conserving energy and resources. This principle may extend to other pathogens and could inform therapeutic strategies aimed at disrupting coenzyme recycling in disease contexts.

The role of polyamines in *Streptococcus pneumoniae* illustrates how small molecules can exert expansive effects on gene regulation, stress tolerance, and virulence. Polyamine metabolism emerges as a potential target for intervention, especially considering rising antibiotic resistance and vaccine limitations. Similarly, the study on sulfate-reducing bacteria shows that microbial metabolites can significantly influence electrochemical processes, with implications for both pathogenesis and bioengineering applications.

Iron metabolism, regulated by Fur in *Riemerella anatipestifer*, exemplifies the delicate balance bacteria must maintain between nutrient acquisition and oxidative stress. Fur's role in modulating iron uptake and ROS resistance directly impacts virulence, reinforcing the idea that metabolic regulation is tightly linked to immune evasion and survival. In *Haemophilus influenzae*, the transcriptional modulation of sialometabolism genes affects surface glycosylation and serum resistance, highlighting the importance of sugar metabolism in host–pathogen interactions.

The global regulator CodY in Gram-positive bacteria serves as a master integrator of nutrient sensing and virulence control. Its ability to change gene expression in response to metabolic status allows pathogens to adapt their behavior to changing environments, facilitating transitions between commensalism and pathogenicity. This regulatory flexibility is mirrored in *Brucella melitensis*, where OtpR coordinates metabolic and virulence gene expression under acidic stress, enabling intracellular survival.

The chapter also emphasizes the metabolic plasticity of intracellular pathogens, such as *Listeria*, *Shigella*, *Salmonella*, and *M. tuberculosis*. These organisms exploit host-derived carbon sources and rewire their metabolic networks to support replication and persistence. The identification of fumarate as a virulence regulator in EHEC further illustrates how central metabolites can influence pathogenic behavior through transcriptional control.

Bacterial responses to environmental stress are multifaceted, involving changes in cell structure, metabolism, and group behavior. The formation of hibernating ribosomes, nucleoprotein complexes, and biofilms reflects the strategic adaptations bacteria employ to conserve energy and resist damage. The discovery that mucin glycans can attenuate *P. aeruginosa* virulence by disrupting biofilms and suppressing quorum sensing opens new avenues for host-derived therapeutic interventions.

Finally, the stringent response, mediated by alarmones such as ppGpp and pGpp, represents a global regulatory system that enables bacteria to survive nutrient deprivation and other stresses. These molecules orchestrate extensive changes in gene expression and metabolism, positioning bacteria for long-term survival and persistence in harsh conditions.

In conclusion, the metabolic pathways explored in this chapter reveal a sophisticated and interconnected network that governs bacterial life. Understanding these systems not only deepens our knowledge of microbial physiology but also offers novel targets for therapeutic intervention. As we continue to face challenges such as antibiotic resistance and emerging pathogens, leveraging insights into bacterial metabolism will be essential for developing innovative strategies to control infection and promote health.

Bibliography

1. Roy, P. H. *et al.* Complete genome sequence of the multiresistant taxonomic outlier Pseudomonas aeruginosa PA7. *PLoS One 5*, e8842 (2010).
2. Balasubramanian, D. & Mathee, K. Comparative transcriptome analyses of Pseudomonas aeruginosa. *Hum. Genomics 3*, 349–361 (2009).
3. Wolfgang, M. C. *et al.* Conservation of genome content and virulence determinants among clinical and environmental isolates of Pseudomonas aeruginosa. *Proc. Natl. Acad. Sci. USA 100*, 8484–8489 (2003).
4. Lee, D. G. *et al.* Genomic analysis reveals that Pseudomonas aeruginosa virulence is combinatorial. *Genome Biol.* 7, R90-2006-7-10-r90. Epub 2006 Oct 12 (2006).

5. Vasquez-Rifo, A., Veksler-Lublinsky, I., Cheng, Z., Ausubel, F. M. & Ambros, V. The Pseudomonas aeruginosa accessory genome elements influence virulence towards Caenorhabditis elegans. *Genome Biol. 20*, 270-019–1890-1 (2019).
6. Apidianakis, Y. & Rahme, L. G. Drosophila melanogaster as a model host for studying Pseudomonas aeruginosa infection. *Nat. Protoc. 4*, 1285–1294 (2009).
7. Yeung, A. T. *et al.* Swarming of Pseudomonas aeruginosa is controlled by a broad spectrum of transcriptional regulators, including MetR. *J. Bacteriol. 191*, 5592–5602 (2009).
8. Tokumoto, U. & Takahashi, Y. Genetic analysis of the *isc* operon in *Escherichia coli* involved in the biogenesis of cellular iron–sulfur proteins. *J. Biochem. 130*, 63–71 (2001).
9. Linster, C. L., Van Schaftingen, E. & Hanson, A. D. Metabolite damage and its repair or pre-emption. *Nat. Chem. Biol. 9*, 72–80 (2013).
10. Kremer, L. S. *et al.* NAXE mutations disrupt the cellular NAD(P)HX repair system and cause a lethal neurometabolic disorder of early childhood. *Am. J. Hum. Genet. 99*, 894–902 (2016).
11. Mills, P. B. *et al.* Mutations in antiquitin in individuals with pyridoxine-dependent seizures. *Nat. Med. 12*, 307–309 (2006).
12. Shah P, Nanduri B, Swiatlo E, Ma Y, Pendarvis K. Polyamine biosynthesis and transport mechanisms are crucial for fitness and pathogenesis of *Streptococcus pneumoniae*. Microbiology. 2010;157:504–15.
13. Ayoola MB, Nakamya MF, Shack LA, Park S, Lim J, Lee JH, et al. SP_0916 is an arginine decarboxylase that catalyzes the synthesis of agmatine, which is critical for capsule biosynthesis in *Streptococcus pneumoniae*. Front Microbiol. 2020;11:578533.
14. Bergel A, Féron D, Mollica A (2005) Catalysis of oxygen reduction in PEM fuel cell by seawater biofilm. Electrochem Commun 7:900–904. https://doi.org/10.1016/j.elecom.2005.06.006
15. De SL, Boeckx P, Verstraete W (2010) Evaluation of biocathodes in freshwater and brackish sediment microbial fuel cells. Appl Microbiol Biotechnol 87:1675–1687. https://doi.org/10.1007/s00253-010-2645-9
16. Guerrini E, Grattieri M, Faggianelli A, Cristiani P, Trasatti S (2015) PTFE effect on the electrocatalysis of the oxygen reduction reaction in membraneless microbial fuel cells. Bioelectrochemistry 106:240–247. https://doi.org/10.1016/j.bioelechem.2015.05.008
17. Yazdi AZ, Roberts EPL, Sundararaj U (2016) Nitrogen/sulfur co-doped helical graphene nanoribbons for efficient oxygen reduction in alkaline and acidic electrolytes. Carbon 100:99–108. https://doi.org/10.1016/j.carbon.2015.12.096
18. Shao M, Chang Q, Dodelet JP, Chenitz R (2016) Recent advances in electrocatalysts for oxygen reduction reaction. Chem Rev 116:3594–3657. https://doi.org/10.1021/acs.chemrev.5b00462
19. Hess C, Enichlmayr H, Jandreski-Cvetkovic D, Liebhart D, Bilic I, Hess M (2013) *Riemerella anatipestifer* outbreaks in commercial goose flocks and identification of isolates by MALDI-TOF mass spectrometry. Avian Pathol 42:151–156. https://doi.org/10.1080/03079457.2013.775401
20. Andrews SC, Robinson AK, Rodriguez-Quinones F (2003) Bacterial iron homeostasis. FEMS Microbiol Rev 27:215–237. https://doi.org/10.1016/S0168-6445(03)00055-X
21. Huynh C, Andrews NW (2008) Iron acquisition within host cells and the pathogenicity of *Leishmania*. Cell Microbiol 10:293–300. https://doi.org/10.1111/j.1462-5822.2007.01095.x
22. Nairz M, Schroll A, Sonnweber T, Weiss G (2010) The struggle for iron - a metal at the host-pathogen interface. Cell Microbiol 12:1691–1702. https://doi.org/10.1111/j.1462-5822.2010.01529.x
23. Chasteen ND, Harrison PM (1999) Mineralization in ferritin: an efficient means of iron storage. J Struct Biol 126:182–194. https://doi.org/10.1006/jsbi.1999.4118
24. Stojiljkovic I, Baumler AJ, Hantke K (1994) Fur regulon in gram-negative bacteria. Identification and characterization of new iron-regulated *Escherichia coli* genes by a fur titration assay. J Mol Biol 236:531–545. https://doi.org/10.1006/jmbi.1994.1163
25. Kurabayashi K, Agata T, Asano H, Tomita H, Hirakawa H (2016) Fur represses adhesion to, invasion of, and intracellular bacterial community formation within bladder epithelial

cells and motility in uropathogenic *Escherichia coli*. Infect Immun 84:3220–3231. https://doi.org/10.1128/IAI.00369-16
26. Hu, Y. H. & Sun, L. The global regulatory effect of *Edwardsiella tarda* Fur on iron acquisition, stress resistance, and host infection: a proteomics-based interpretation. *J. Proteom. 140*, 100–110 (2016).
27. Choi J, Ryu S (2019) Regulation of iron uptake by fine-tuning the iron responsiveness of the iron sensor Fur. Appl Environ Microbiol 85:e03026–e3118. https://doi.org/10.1128/AEM.03026-18
28. Santos R, Batista BB, da Silva Neto JF (2020) Ferric uptake regulator Fur coordinates siderophore production and defense against iron toxicity and oxidative stress and contributes to virulence in *Chromobacterium violaceum*. Appl Environ Microbiol 86:e01620–e1720. https://doi.org/10.1128/AEM.01620-20
29. Liao H, Cheng X, Zhu D, Wang M, Jia R, Chen S, Chen X, Biville F, Liu M, Cheng A (2015) TonB Energy transduction systems of *Riemerella anatipestifer* are required for iron and hemin utilization. PLoS One 10:e0127506. https://doi.org/10.1371/journal.pone.0127506
30. Liu M, Wang M, Zhu D, Wang M, Jia R, Chen S, Sun K, Yang Q, Wu Y, Chen X, Biville F, Cheng A (2016) Investigation of TbfA in *Riemerella anatipestifer* using plasmid-based methods for gene over-expression and knockdown. Sci Rep 6:37159. https://doi.org/10.1038/srep37159
31. Wang X, Liu W, Zhu D, Yang L, Liu M, Yin S, Wang M, Jia R, Chen S, Sun K, Cheng A, Chen X (2014) Comparative genomics of *Riemerella anatipestifer* reveals genetic diversity. BMC Genomics 15:479. https://doi.org/10.1186/1471-2164-15-479
32. Guo Y, Hu D, Guo J, Li X, Guo J, Wang X, Xiao Y, Jin H, Liu M, Li Z, Bi D, Zhou Z (2017) The role of the regulator Fur in gene regulation and virulence of *Riemerella anatipestifer* assessed using an unmarked gene deletion system. Front Cell Infect Microbiol 7:382. https://doi.org/10.3389/fcimb.2017.00382
33. Crosa LM, Crosa JH, Heffron F (2009) Iron transport in *Francisella* in the absence of a recognizable TonB protein still requires energy generated by the proton motive force. Biometals 22:337–344. https://doi.org/10.1007/s10534-008-9170-7
34. Yeowell HN, White JR (1982) Iron requirement in the bactericidal mechanism of streptonigrin. Antimicrob Agents Chemother 22:961–968
35. Liu M, Huang M, Zhu D, Wang M, Jia R, Chen S, Sun K, Yang Q, Wu Y, Biville F, Cheng A (2017) Identifying the genes responsible for iron-limited condition in *Riemerella anatipestifer* CH-1 through RNA-Seq-Based analysis. Biomed Res Int 2017:8682057. https://doi.org/10.1155/2017/8682057
36. Hood DW, Makepeace K, Deadman ME, Rest RF, Thibault P, Martin A, Richards JC, Moxon ER: Sialic acid in the lipopolysaccharide of Haemophilus influenzae: strain distribution, influence on serum resistance and structural characterization. Mol Microbiol. 1999, 33 (4): 679–692. https://doi.org/10.1046/j.1365-2958.1999.01509.x.
37. Vimr E, Lichtensteiger C, Steenbergen S: Sialic acid metabolism's dual function in Haemophilus influenzae. Mol Microbiol. 2000, 36 (5): 1113–1123. https://doi.org/10.1046/j.1365-2958.2000.01925.x.
38. Severi E, Randle G, Kivlin P, Whitfield K, Young R, Moxon R, Kelly D, Hood D, Thomas GH: Sialic acid transport in Haemophilus influenzae is essential for lipopolysaccharide sialylation and serum resistance and is dependent on a novel tripartite ATP-independent periplasmic transporter. Mol Microbiol. 2005, 58 (4): 1173–1185. https://doi.org/10.1111/j.1365-2958.2005.04901.x.
39. Allen S, Zaleski A, Johnston JW, Gibson BW, Apicella MA: Novel sialic acid transporter of Haemophilus influenzae. Infect Immun. 2005, 73 (9): 5291–5300. https://doi.org/10.1128/IAI.73.9.5291-5300.2005.
40. Johnston JW, Zaleski A, Allen S, Mootz JM, Armbruster D, Gibson BW, Apicella MA, Munson RS: Regulation of sialic acid transport and catabolism in Haemophilus influenzae. Mol Microbiol. 2007, 66 (1): 26–39. https://doi.org/10.1111/j.1365-2958.2007.05890.x.

41. Herriott RM, Meyer EM, Vogt M: Defined nongrowth media for stage II development of competence in Haemophilus influenzae. J Bacteriol. 1970, 101 (2): 517–524.
42. Grundy FJ, Henkin TM (1993) tRNA as a positive regulator of transcription antitermination in *B. subtilis*. Cell 74:475–482
43. Grundy FJ, Henkin TM (1998) The S box regulon: a new global transcription termination control system for methionine and cysteine biosynthesis genes in gram-positive bacteria. Mol Microbiol 30:737–749
44. Serganov A et al (2004) Structural basis for discriminative regulation of gene expression by adenine- and guanine-sensing mRNAs. Chem Biol 11:1729–1741
45. Ababneh QO, Herman JK (2015) CodY regulates SigD levels and activity by binding to three sites in the *fla/che* operon. J Bacteriol 197:2999–3006
46. Brinsmade S, Alexander E, Livny J, Stettner A, Segrè D, Rhee K, Sonenshein A (2014) Hierarchical expression of genes controlled by the *Bacillus subtilis* global regulatory protein CodY. Proc Natl Acad Sci USA 111:8227–8232
47. Molle V, Nakaura Y, Shivers RP, Yamaguchi H, Losick R, Fujita Y, Sonenshein AL (2003) Additional targets of the *Bacillus subtilis* global regulator CodY identified by chromatin immunoprecipitation and genome-wide transcript analysis. J Bacteriol 185:1911–1922
48. Shivers RP, Dineen SS, Sonenshein AL (2006) Positive regulation of *Bacillus subtilis ackA* by CodY and CcpA: establishing a potential hierarchy in carbon flow. Mol Microbiol 62:811–822
49. Belitsky BR, Sonenshein AL (2008) Genetic and biochemical analysis of CodY-binding sites in *Bacillus subtilis*. J Bacteriol 190:1224–1236
50. Brinsmade SR, Sonenshein AL (2011) Dissecting complex metabolic integration provides direct genetic evidence for CodY activation by guanine nucleotides. J Bacteriol 193:5637–5648
51. Shivers RP, Sonenshein AL (2004) Activation of the *Bacillus subtilis* global regulator CodY by direct interaction with branched-chain amino acids. Mol Microbiol 53:599–611
52. Ibberson C, Jones C, Singh S, Wise M, Hart M, Zurawski D, Horswill A (2014) *Staphylococcus aureus* hyaluronidase is a CodY-regulated virulence factor. Infect Immun 82:4253–4264
53. Pohl K et al (2009) CodY in *Staphylococcus aureus*: a regulatory link between metabolism and virulence gene expression. J Bacteriol 191:2953–2963
54. Waters NR, Samuels DJ, Behera RK, Livny J, Rhee KY, Sadykov MR, Brinsmade SR (2016) A spectrum of CodY activities drives metabolic reorganization and virulence gene expression in *Staphylococcus aureus*. Mol Microbiol 101:495–514
55. Lobel L, Sigal N, Borovok I, Ruppin E, Herskovits AA (2012) Integrative genomic analysis identifies isoleucine and CodY as regulators of *Listeria monocytogenes* virulence. PLoS Genet 8:e1002887
56. Lobel L, Sigal N, Borovok I, Belitsky BR, Sonenshein AL, Herskovits AA (2015) The metabolic regulator CodY links *Listeria monocytogenes* metabolism to virulence by directly activating the virulence regulatory gene *prfA*. Mol Microbiol 95:624–644
57. Belitsky BR, Sonenshein AL (2011a) Contributions of multiple binding sites and effector-independent binding to CodY-mediated regulation in *Bacillus subtilis*. J Bacteriol 193:473–484
58. Villapakkam AC, Handke LD, Belitsky BR, Levdikov VM, Wilkinson AJ, Sonenshein AL (2009) Genetic and biochemical analysis of the interaction of *Bacillus subtilis* CodY with branched-chain amino acids. J Bacteriol 191:6865–6876
59. Dineen SS, Villapakkam AC, Nordman JT, Sonenshein AL (2007) Repression of *Clostridium difficile* toxin gene expression by CodY. Mol Microbiol 66:206–219
60. Dineen SS, McBride SM, Sonenshein AL (2010) Integration of metabolism and virulence by *Clostridium difficile* CodY. J Bacteriol 192:5350–5362
61. Majerczyk CD, Sadykov MR, Luong TT, Lee C, Somerville GA, Sonenshein AL (2008) *Staphylococcus aureus* CodY negatively regulates virulence gene expression. J Bacteriol 190:2257–2265

62. Majerczyk CD et al (2010) Direct targets of CodY in *Staphylococcus aureus*. J Bacteriol 192:2861–2877
63. Grosser MR, Weiss A, Shaw LN, Richardson AR (2016) Regulatory requirements for *Staphylococcus aureus* nitric oxide resistance. J Bacteriol 198:2043–2055
64. Greene C, McDevitt D, Francois P, Vaudaux P, Lew D, Foster T (1995) Adhesion properties of mutants of *Staphylococcus aureus* defective in fibronectin-binding proteins and studies on the expression of *fnb* genes. Mol Microbiol 17:1143–1152
65. Hartleib J et al (2000) Protein A is the von Willebrand factor binding protein on *Staphylococcus aureus*. Blood 96:2149–2156
66. George SE, Nguyen T, Geiger T, Weidenmaier C, Lee JC, Liese J, Wolz C (2015) Phenotypic heterogeneity and temporal expression of the capsular polysaccharide in *Staphylococcus aureus*. Mol Microbiol 98:1073–1088
67. Luong T, Sau S, Gomez M, Lee JC, Lee CY (2002) Regulation of *Staphylococcus aureus* capsular polysaccharide expression by *agr* and *sarA*. Infect Immun 70:444–450
68. Zhang, X., Ren, J., Li, N., Liu, W. & Wu, Q. Disruption of the BMEI0066 gene attenuates the virulence of Brucella melitensis and decreases its stress tolerance. Int J Biol Sci 5, 570–577 (2009).
69. Liu, W. et al. OtpR regulated the growth, cell morphology of B. melitensis and tolerance to beta-lactam agents. Vet Microbiol 159, 90–98 (2012).
70. Roop, R. M., 2nd, Gaines, J. M., Anderson, E. S., Caswell, C. C. & Martin, D. W. Survival of the fittest: how Brucella strains adapt to their intracellular niche in the host. Med Microbiol Immunol 198, 221–238 (2009).
71. Atluri, V. L., Xavier, M. N., de Jong, M. F., den Hartigh, A. B. & Tsolis, R. M. Interactions of the human pathogenic Brucella species with their hosts. Annu Rev Microbiol 65, 523–541 (2011).
72. Arenas, G. N., Staskevich, A. S., Aballay, A. & Mayorga, L. S. Intracellular trafficking of Brucella abortus in J774 macrophages. Infect Immun 68, 4255–4263 (2000).
73. Porte, F., Liautard, J. P. & Kohler, S. Early acidification of phagosomes containing Brucella suis is essential for intracellular survival in murine macrophages. Infect Immun 67, 4041–4047. (1999).
74. Alpuche Aranda, C. M., Swanson, J. A., Loomis, W. P. & Miller, S. I. Salmonella typhimurium activates virulence gene transcription within acidified macrophage phagosomes. Proc Natl Acad Sci U S A 89, 10079–10083 (1992).
75. Mortazavi, A., Williams, B. A., McCue, K., Schaeffer, L. & Wold, B. Mapping and quantifying mammalian transcriptomes by RNA-Seq. Nature methods 5, 621–628 (2008).
76. Chen, J. C. & Beckwith, J. FtsQ, FtsL and FtsI require FtsK, but not FtsN, for co-localization with FtsZ during Escherichia coli cell division. Mol Microbiol 42, 395–413 (2001).
77. Rambow-Larsen, A. A., Petersen, E. M., Gourley, C. R. & Splitter, G.A. Brucella regulators: self-control in a hostile environment. Trends Microbiol 17, 371–377 (2009).
78. Fugier, E., Pappas, G. & Gorvel, J. P. Virulence factors in brucellosis: implications for aetiopathogenesis and treatment. Expert Rev Mol Med 9, 1–10 (2007).
79. Xiang, Z., Tian, Y. & He, Y. PHIDIAS: a pathogen-host interaction data integration and analysis system. Genome Biol 8, R150 (2007).
80. He, Y. Analyses of Brucella Pathogenesis, Host Immunity and Vaccine Targets using Systems Biology and Bioinformatics. Front. Cell. Infect. Microbiol 2, 2 (2012).
81. Wright, E. M. et al. Active sugar transport in eukaryotes. *J. Exp. Biol. 196*, 197–212 (1994).
82. Munger, J. et al. Systems-level metabolic flux profiling identifies fatty acid synthesis as a target for antiviral therapy. *Nature Biotech. 26*, 1179–1186 (2008).
83. Nielsen, J. & Oliver, S. The next wave in metabolome analysis. *Trends Biotechnol. 23*, 544–546 (2005).
84. Nicholson, J. K & Lindon, J. C. Metabonomics. *Nature 455*, 1054–1056 (2008).
85. Breitling, R., Vitkup, D. & Barrett, M. P. New surveyor tools for charting microbial metabolic maps. *Nature Rev. Microbiol. 6*, 156–161 (2008).

86. Swain, R. J. & Stevens, M. M. Raman microspectroscopy for non-invasive biochemical analysis of single cells. *Biochem. Soc. Trans. 35*, 544–549 (2007).
87. Wagner, M., Single-cell ecophysiology of microbes as revealed by Raman microspectroscopy or secondary ion mass spectrometry imaging. *Ann. Rev. Microbiol. 63*, 411–429 (2009).
88. Melican, K., Richter-Dahlfors, A. Real-time live imaging to study bacterial infections *in vivo*. *Curr. Opin. Microbiol. 12*, 31–36 (2009).
89. Bumann, D. & Valdivia, R. H. Identification of host-induced pathogen genes by differential fluorescence induction reporter systems. *Nature Protoc. 2*, 770–777 (2007).
90. Runyen-Janecky, L. J. & Payne, S. M. Identification of chromosomal *Shigella flexneri* genes induced by the eukaryotic intracellular environment. *Infect. Immun. 70*, 4379–4388 (2002).
91. Eriksson, S., Lucchini, S., Thompson, A., Rhen, M. & Hinton, J. C. Unravelling the biology of macrophage infection by gene expression profiling of intracellular *Salmonella enterica*. *Mol. Microbiol. 47*, 103–118 (2003).
92. Schnappinger, D. et al. Transcriptional adaptation of *Mycobacterium tuberculosis* within macrophages: insights into the phagosomal environment. *J. Exp. Med. 198*, 693–704 (2003). *The first comprehensive transcriptome analysis of M. tuberculosis in resting and activated macrophages, with a discussion of the metabolic implications.*
93. Lucchini, S., Liu, H., Jin, Q., Hinton, J. C. & Yu, J. Transcriptional adaptation of *Shigella flexneri* during infection of macrophages and epithelial cells: insights into the strategies of a cytosolic bacterial pathogen. *Infect. Immun. 73*, 88–102 (2005). *The first comprehensive comparative transcriptome analysis of intracellularly replicating S. flexneri.*
94. Joseph, B. et al. Identification of *Listeria monocytogenes* genes contributing to intracellular replication by expression profiling and mutant screening. *J. Bacteriol. 188*, 556–568 (2006).
95. Chatterjee, S. S. et al. Intracellular gene expression profile of *Listeria monocytogenes*. *Infect. Immun. 74*, 1323–1338 (2006). *This article and reference 73 show the first comprehensive transcriptome analyses of intracellularly replicating L. monocytogenes.*
96. Hautefort, I. et al. During infection of epithelial cells *Salmonella enterica* serovar Typhimurium undergoes a time-dependent transcriptional adaptation that results in simultaneous expression of three type 3 secretion systems. *Cell. Microbiol. 10*, 958–984 (2008). *A thorough transcriptome analysis of intracellularly replicating S. enterica.*
97. Chico-Calero, I. et al. Hpt, a bacterial homolog of the microsomal glucose-6-phosphate translocase, mediates rapid intracellular proliferation in *Listeria*. *Proc. Natl Acad. Sci. USA 99*, 431–436 (2002). *This article demonstrates that a metabolic gene partially needed for the metabolism of intracellular L. monocytogenes is co-regulated with other virulence genes.*
98. Eylert, E. et al. Carbon metabolism of *Listeria monocytogenes* growing inside macrophages. *Mol. Microbiol. 69*, 1008–1017 (2008). *The first application of the* <Superscript>*13*</Superscript>*C-IPA method for studying the carbon metabolism of an intracellular pathogen in host cells.*
99. Goetz, A., Eylert, E., Eisenreich, W. & Goebel, W. Carbon metabolism of enterobacterial human pathogens growing in epithelial colorectal adenocarcinoma (Caco-2) cells. *PLoS ONE* (in the press).
100. Hernandez-Doria, J. D. & Sperandio, V. Nutrient and chemical sensing by intestinal pathogens. *Microbes Infect. 15*, 759–764 (2013).
101. Rohmer, L., Hocquet, D. & Miller, S. I. Are pathogenic bacteria just looking for food? Metabolism and microbial pathogenesis. *Trends Microbiol. 19*, 341–348 (2011).
102. Barnett Foster, D. Modulation of the enterohemorrhagic *E. coli* virulence program through the human gastrointestinal tract. *Virulence 4*, 315–323 (2013).
103. Keeney, K. M. & Finlay, B. B. Enteric pathogen exploitation of the microbiota-generated nutrient environment of the gut. *Curr. Opin. Microbiol. 14*, 92–98 (2011).
104. Njoroge, J. W., Nguyen, Y., Curtis, M. M., Moreira, C. G. & Sperandio, V. Virulence meets metabolism: Cra and KdpE gene regulation in enterohemorrhagic *Escherichia coli*. *mBio 3*, e00280-12 (2012).

105. Chaba R, Alba BM, Guo MS et al (2011) Signal integration by DegS and RseB governs the σ^E-mediated envelope stress response in *Escherichia coli*. Proc Natl Acad Sci USA 108:2106–2111. https://doi.org/10.1073/pnas.1019277108
106. Kim DY (2015) Two stress sensor proteins for the expression of sigmaE regulon: DegS and RseB. J Microbiol 53:306–310. https://doi.org/10.1007/s12275-015-5112-6
107. Aertsen A, Michiels CW (2004) Stress and how bacteria cope with death and survival. Crit Rev Microbiol 30:263–273. https://doi.org/10.1080/10408410490884757
108. Finkel SE (2006) Long-term survival during stationary phase: evolution and the GASP phenotype. Nat Rev Microbiol 4:113–120. https://doi.org/10.1038/nrmicro1340
109. Goswami G, Panda D, Samanta R et al (2018) *Bacillus megaterium* adapts to acid stress condition through a network of genes: insight from a genome-wide transcriptome analysis. Sci Rep 8:16105. https://doi.org/10.1038/s41598-018-34221-0
110. Nikaido H (2003) Molecular basis of bacterial outer membrane permeability revisited. Microbiol Mol Biol Rev 67:593–656. https://doi.org/10.1128/mmbr.67.4.593-656.2003
111. Silhavy TJ, Kahne D, Walker S (2010) The bacterial cell envelope. Cold Spring Harb Perspect Biol 2:a000414. https://doi.org/10.1101/cshperspect.a000414
112. Lee DD, Galera-Laporta L, Bialecka-Fornal M et al (2019) Magnesium flux modulates ribosomes to increase bacterial survival. Cell 177:352–360. https://doi.org/10.1016/j.cell.2019.01.042
113. Mullins EA, Kappock TJ (2012) Crystal structures of *Acetobacter aceti* succinyl-coenzyme A (CoA): acetate CoA-transferase reveal specificity determinants and illustrate the mechanism used by class I CoA-transferases. Biochemistry 51:8422–8434. https://doi.org/10.1021/bi300957f
114. Kwong WK, Zheng H, Moran NA (2017) Convergent evolution of a modified, acetate-driven TCA cycle in bacteria. Nat Microbiol 2:17067–17067. https://doi.org/10.1038/nmicrobiol.2017.67
115. Shimizu K (2013) Regulation systems of bacteria such as *Escherichia coli* in response to nutrient limitation and environmental stresses. Metabolites 4:1–35. https://doi.org/10.3390/metabo4010001
116. Lund P, Tramonti A, De Biase D (2015) Coping with low pH: molecular strategies in neutralophilic bacteria. FEMS Microbiol Rev 38:1091–1125. https://doi.org/10.1111/1574-6976.12076
117. Aquino P, Honda B, Jaini S et al (2017) Coordinated regulation of acid resistance in *Escherichia coli*. BMC Syst Biol 11:1. https://doi.org/10.1186/s12918-016-0376-y
118. Shrout J, Chopp DL, Just CL et al (2010) The impact of quorum sensing and swarming motility on *Pseudomonas aeruginosa* biofilm formation is nutritionally conditional. Mol Microbiol 62:1264–1277. https://doi.org/10.1111/j.1365-2958.2006.05421.x
119. Wei Y, Vollmer AC, LaRossa RA (2001) In vivo titration of mitomycin C action by four *Escherichia coli* genomic regions on multicopy plasmids. J Bacteriol 183:2259–2264. https://doi.org/10.1128/JB.183.7.2259-2264.2001
120. Wu C, Huang J, Zhou R (2014) Progress in engineering acid stress resistance of lactic acid bacteria. Appl Microbiol Biotechnol 98:1055–1063. https://doi.org/10.1007/s00253-013-5435-3
121. Lindemann SR, Bernstein HC, Song HS et al (2016) Engineering microbial consortia for controllable outputs. ISME J 10:2077–2084. https://doi.org/10.1038/ismej.2016.26
122. Schuster M, Hawkins AC, Harwood CS et al (2004) The *Pseudomonas aeruginosa* RpoS regulon and its relationship to quorum sensing. Mol Microbiol 51:973–985. https://doi.org/10.1046/j.1365-2958.2003.03886.x
123. Chua, S. L. et al. In vitro and in vivo generation and characterization of *Pseudomonas aeruginosa* biofilm-dispersed cells via c-di-GMP manipulation. *Nat. Protoc. 10*, 1165–1180 (2015).
124. Griffin, J. E. et al. High-resolution phenotypic profiling defines genes essential for mycobacterial growth and cholesterol catabolism. *PLoS Pathog. 7*, e1002251 (2011).
125. El-Gebali, S. et al. The Pfam protein families database in 2019. *Nucleic Acids Res. 47*, D427–D432 (2019).

126. Mitchell, A. L. et al. InterPro in 2019: improving coverage, classification and access to protein sequence annotations. *Nucleic Acids Res. 47*, D351–D360 (2019).
127. Huerta-Cepas, J. et al. eggNOG 4.5: a hierarchical orthology framework with improved functional annotations for eukaryotic, prokaryotic and viral sequences. *Nucleic Acids Res. 44*, D286–D293 (2016).
128. Yang, Z., Zeng, X. & Tsui, S. K.-W. Investigating function roles of hypothetical proteins encoded by the *Mycobacterium tuberculosis* H37Rv genome. *BMC Genomics 20*, 394 (2019).
129. Ortega, C. *et al.* Systematic survey of serine hydrolase activity in *Mycobacterium tuberculosis* defines changes associated with persistence. *Cell Chem. Biol. 23*, 290–298 (2016).
130. Penn, B. H. et al. An Mtb-human protein-protein interaction map identifies a switch between host antiviral and antibacterial responses. *Mol. Cell 71*, 637–648.e5 (2018).
131. Wang, M., Herrmann, C. J., Simonovic, M., Szklarczyk, D. & von Mering, C. Version 4.0 of PaxDb: protein abundance data, integrated across model organisms, tissues, and cell-lines. *Proteomics 15*, 3163–3168 (2015).
132. Boutte, C. C. & Crosson, S. Bacterial lifestyle shapes stringent response activation. *Trends Microbiol. 21*, 174–180 (2013).
133. Potrykus, K. & Cashel, M. (p)ppGpp: still magical? *Annu. Rev. Microbiol. 62*, 35–51 (2008).
134. Ronneau, S. & Hallez, R. Make and break the alarmone: regulation of (p)ppGpp synthetase/hydrolase enzymes in bacteria. *FEMS Microbiol. Rev. 43*, 389–400 (2019).
135. Irving, S. E. & Corrigan, R. M. Triggering the stringent response: signals responsible for activating (p)ppGpp synthesis in bacteria. *Microbiology 164*, 268–276 (2018).
136. Sy, J. *In vitro* degradation of guanosine 5′-diphosphate, 3′-diphosphate. *Proc. Natl Acad. Sci. USA 74*, 5529–5533 (1977).
137. Aravind, L. & Koonin, E. V. The HD domain defines a new superfamily of metal-dependent phosphohydrolases. *Trends Biochem. Sci. 23*, 469–472 (1998).
138. Keasling, J. D., Bertsch, L. & Kornberg, A. Guanosine pentaphosphate phosphohydrolase of *Escherichia coli* is a long-chain exopolyphosphatase. *Proc. Natl Acad. Sci. USA 90*, 7029–7033 (1993).
139. Gaca, A. O. et al. From (p)ppGpp to (pp)pGpp: characterization of regulatory effects of pGpp synthesized by the small alarmone synthetase of *Enterococcus faecalis*. *J. Bacteriol. 197*, 2908–2919 (2015).
140. Travers, A. A. ppApp alters transcriptional selectivity of *Escherichia coli* RNA polymerase. *FEBS Lett. 94*, 345–348 (1978).
141. Bruhn-Olszewska, B. et al. Structure-function comparisons of (p)ppApp vs (p)ppGpp for *Escherichia coli* RNA polymerase binding sites and for *rrnB* P1 promoter regulatory responses vitro. *Biochim. Biophys. Acta Gene Regul. Mech. 1861*, 731–742 (2018).
142. Ahmad, S. et al. An interbacterial toxin inhibits target cell growth by synthesizing (p)ppApp. *Nature 575*, 674–678 (2019). *This article shows the role of signalling nucleotides as toxins.*
143. Wang, J. D., Sanders, G. M. & Grossman, A. D. Nutritional control of elongation of DNA replication by (p)ppGpp. *Cell 128*, 865–875 (2007).
144. Maciag, M., Kochanowska, M., Lyzen, R., Wegrzyn, G. & Szalewska-Palasz, A. ppGpp inhibits the activity of *Escherichia coli* DnaG primase. *Plasmid 63*, 61–67 (2010).
145. Rymer, R. U. et al. Binding mechanism of metalNTP substrates and stringent-response alarmones to bacterial DnaG-type primases. *Structure 20*, 1478–1489 (2012).
146. Wang, B., Grant, R. A. & Laub, M. T. ppGpp coordinates nucleotide and amino-acid synthesis in *E. coli* during starvation. *Mol. Cell* https://doi.org/10.1016/j.molcel.2020.08.005 (2020).
147. Wang, B. et al. Affinity-based capture and identification of protein effectors of the growth regulator ppGpp. *Nat. Chem. Biol. 15*, 141–150 (2019). *This work uses a capture compound to expand the repertoire of (p)ppGpp target proteins.*
148. Battesti, A. & Bouveret, E. Acyl carrier protein/SpoT interaction, the switch linking SpoT-dependent stress response to fatty acid metabolism. *Mol. Microbiol. 62*, 1048–1063 (2006).
149. Gully, D., Moinier, D., Loiseau, L. & Bouveret, E. New partners of acyl carrier protein detected in *Escherichia coli* by tandem affinity purification. *FEBS Lett. 548*, 90–96 (2003).

150. Butland, G. et al. Interaction network containing conserved and essential protein complexes in *Escherichia coli*. *Nature 433*, 531–537 (2005).
151. Sinha, A. K., Winther, K. S., Roghanian, M. & Gerdes, K. Fatty acid starvation activates RelA by depleting lysine precursor pyruvate. *Mol. Microbiol. 112*, 1339–1349 (2019).
152. Fitzsimmons, L. F. et al. SpoT induces intracellular *Salmonella* virulence programs in the phagosome. *mBio 11*, e03397-19 (2020).
153. Gaynor, E. C., Wells, D. H., MacKichan, J. K. & Falkow, S. The *Campylobacter jejuni* stringent response controls specific stress survival and virulence-associated phenotypes. *Mol. Microbiol. 56*, 8–27 (2005).
154. Zhu, J. et al. (p)ppGpp synthetases regulate the pathogenesis of zoonotic *Streptococcus suis*. *Microbiol. Res. 191*, 1–11 (2016).
155. Hammer, B. K. & Swanson, M. S. Co-ordination of *Legionella pneumophila* virulence with entry into stationary phase by ppGpp. *Mol. Microbiol. 33*, 721–731 (1999).
156. Colomer-Winter, C., Gaca, A. O., Chuang-Smith, O. N., Lemos, J. A. & Frank, K. L. Basal levels of (p)ppGpp differentially affect the pathogenesis of infective endocarditis in *Enterococcus faecalis*. *Microbiology 164*, 1254–1265 (2018).
157. Colomer-Winter, C., Flores-Mireles, A. L., Kundra, S., Hultgren, S. J. & Lemos, J. A. (p) ppGpp and CodY promote *Enterococcus faecalis* virulence in a murine model of catheter-associated urinary tract infection. *mSphere 4*, e00392-19 (2019).
158. Jimmy, S. et al. A widespread toxin-antitoxin system exploiting growth control via alarmone signaling. *Proc. Natl Acad. Sci. USA 117*, 10500–10510 (2020).
159. Panayidou, Stavria; Georgiades, Kaliopi; Christofi, Theodoulakis; Tamana, Stella; Promponas, Vasilis J.; Apidianakis, Yiorgos. Pseudomonas aeruginosa core metabolism exerts a widespread growth-independent control on virulence. Scientific Reports (2020). https://doi.org/10.1038/s41598-020-66194-4
160. Hartl, Johannes; Kiefer, Patrick; Meyer, Fabian; Vorholt, Julia A. Longevity of major coenzymes allows minimal de novo synthesis in microorganisms. Nature Microbiology (2017). https://doi.org/10.1038/nmicrobiol.2017.73
161. Nanduri, Bindu; Swiatlo, Edwin. The expansive effects of polyamines on the metabolism and virulence of Streptococcus pneumoniae. Pneumonia (2021). https://doi.org/10.1186/s41479-021-00082-x
162. Wu, Jiajia; Liu, Huaiqun; Wang, Peng; Zhang, Dun; Sun, Yan; Li, Ee. Oxygen Reduction Reaction Affected by Sulfate-Reducing Bacteria: Different Roles of Bacterial Cells and Metabolites. Indian Journal of Microbiology (2017). https://doi.org/10.1007/s12088-017-0667-z
163. Huang, Mi; Liu, Mafeng; Liu, Jiajun; Zhu, Dekang; Tang, Qianying; Jia, Renyong; Chen, Shun; Zhao, Xinxin; Yang, Qiao; Wu, Ying; Zhang, Shaqiu; Huang, Juan; Ou, Xumin; Mao, Sai; Gao, Qun; Sun, Di; Wang, Mingshu; Cheng, Anchun Functional characterization of Fur in iron metabolism, oxidative stress resistance and virulence of Riemerella anatipestifer. Veterinary Research (2021). https://doi.org/10.1186/s13567-021-00919-9
164. Jenkins, Gaynor A; Figueira, Marisol; Kumar, Gaurav A; Sweetman, Wendy A; Makepeace, Katherine; Pelton, Stephen I; Moxon, Richard; Hood, Derek W Sialic acid mediated transcriptional modulation of a highly conserved sialometabolism gene cluster in Haemophilus influenzae and its effect on virulence. BMC Microbiology (2010). https://doi.org/10.1186/1471-2180-10-48
165. Brinsmade, Shaun R. CodY, a master integrator of metabolism and virulence in Gram-positive bacteria. Current Genetics (2016). https://doi.org/10.1007/s00294-016-0656-5
166. Liu, Wenxiao; Dong, Hao; Li, Jing; Ou, Qixing; Lv, Yujin; Wang, Xiaolei; Xiang, Zuoshuang; He, Yongqun; Wu, Qingmin RNA-seq reveals the critical role of OtpR in regulating Brucella melitensis metabolism and virulence under acidic stress. Scientific Reports (2015). https://doi.org/10.1038/srep10864

167. Eisenreich, Wolfgang; Dandekar, Thomas; Heesemann, Jürgen; Goebel, Werner. Carbon metabolism of intracellular bacterial pathogens and possible links to virulence. Nature Reviews Microbiology (2010). https://doi.org/10.1038/nrmicro2351
168. Kuo, Cheng-Ju; Wang, Sin-Tian; Lin, Chia-Mei; Chiu, Hao-Chieh; Huang, Cheng-Rung; Lee, Der-Yen; Chang, Geen-Dong; Chou, Ting-Chen; Chen, Jenn-Wei; Chen, Chang-Shi. A multi-omic analysis reveals the role of fumarate in regulating the virulence of enterohemorrhagic Escherichia coli. Cell Death & Disease (2018). https://doi.org/10.1038/s41419-018-0423-2
169. Zhang, Xinyi; Li, Zhendong; Pang, Shengmei; Jiang, Boyu; Yang, Yang; Duan, Qiangde; Zhu, Guoqiang The impact of cell structure, metabolism and group behavior for the survival of bacteria under stress conditions. Archives of Microbiology (2020). https://doi.org/10.1007/s00203-020-02050-3
170. Wheeler, Kelsey M.; Cárcamo-Oyarce, Gerardo; Turner, Bradley S.; Dellos-Nolan, Sheri; Co, Julia Y.; Lehoux, Sylvain; Cummings, Richard D.; Wozniak, Daniel J.; Ribbeck, Katharina Mucin glycans attenuate the virulence of Pseudomonas aeruginosa in infection. Nature Microbiology (2019). https://doi.org/10.1038/s41564-019-0581-8
171. Jansen, Robert S.; Mandyoli, Lungelo; Hughes, Ryan; Wakabayashi, Shoko; Pinkham, Jessica T.; Selbach, Bruna; Guinn, Kristine M.; Rubin, Eric J.; Sacchettini, James C.; Rhee, Kyu Y. Aspartate aminotransferase Rv3722c governs aspartate-dependent nitrogen metabolism in Mycobacterium tuberculosis. Nature Communications (2020). https://doi.org/10.1038/s41467-020-15876-8
172. Irving, Sophie E.; Choudhury, Naznin R.; Corrigan, Rebecca M. The stringent response and physiological roles of (pp)pGpp in bacteria. Nature Reviews Microbiology (2020). https://doi.org/10.1038/s41579-020-00470-y

Chapter 3
Methods for Studying Bacterial Metabolism

Vijay Soni and Aditya Upadhyay

3.1 Introduction by the Editor

Understanding bacterial metabolism is central to unraveling the complexities of microbial physiology, pathogenesis, and ecological interactions. As the field of microbiology evolves, so too do the tools and techniques used to study bacterial metabolic processes. This chapter focuses on the diverse and innovative methodologies that have emerged to investigate bacterial metabolism, ranging from computational modeling and high-throughput metabolomics to real-time sensing technologies and microscale platforms. These approaches not only enhance our ability to profile metabolic activity but also enable us to predict, monitor, and manipulate microbial behavior in both clinical and environmental contexts.

One of the most transformative developments in recent years is the integration of metagenomic and metabolomic data. The MelonnPan framework exemplifies this by using machine learning to predict metabolite profiles from microbial community sequencing data. This approach circumvents the need for direct metabolomic profiling, offering a cost-effective and scalable solution for studying microbiomes in health and disease. Similarly, genome-resolved metagenomics and metatranscriptomics have been applied to thermophilic composting systems, revealing key bacterial players and their metabolic interactions. These studies underscore the power of multi-omic integration in capturing the functional dynamics of complex microbial ecosystems.

Intracellular pathogens such as *Legionella pneumophila* and *Coxiella burnetii* present unique challenges for metabolic studies due to their reliance on host cell

V. Soni (✉)
Department of Medicine, Weill Cornell Medicine, New York, NY, USA
e-mail: vis2032@med.cornell.edu

A. Upadhyay
Hospital for Special Surgery, New York, NY, USA

V. Soni, A. Upadhyay (eds.), *Metabolism of Bacterial Infection*,
https://doi.org/10.1007/978-3-031-46895-7_3

machinery. Investigations into their Dot/Icm secretion systems have shed light on how bacterial effectors manipulate host lipid metabolism, vesicle trafficking, and gene expression to sustain intracellular replication. These insights not only deepen our understanding of host–pathogen interactions but also highlight the metabolic dependencies that could be targeted for therapeutic intervention.

Analytical chemistry has also played a pivotal role in advancing bacterial metabolomics. The development of high-performance liquid chromatography (HPLC) methods for detecting phytoestrogen transformations by gut bacteria illustrates how precise analytical tools can uncover the metabolic contributions of microbiota to host physiology. Similarly, redox-reactive nanosensor arrays based on silicon nanowire field-effect transistors (SiNW-FETs) offer real-time, label-free monitoring of bacterial biofilm metabolism. These technologies provide unprecedented sensitivity and temporal resolution, enabling researchers to observe metabolic changes before they manifest in phenotypic alterations.

Rapid diagnostics are another frontier where metabolomics is making a significant impact. The metabolic preference assay (MPA) and metabolic inhibition assay (MIA) represent novel strategies for identifying bloodstream infection pathogens and assessing antimicrobial susceptibility. By analyzing patterns of metabolite consumption and excretion, these assays dramatically reduce diagnostic timelines, improving patient outcomes and supporting antibiotic stewardship.

Microscale platforms have revolutionized the study of microbial secondary metabolism, particularly in filamentous fungi. Open microfluidic systems allow controlled culturing and solvent extraction, facilitating the discovery of bioactive compounds under diverse environmental conditions. These platforms are especially valuable for coculture experiments, where microbial interactions can trigger the production of novel metabolites.

Computational tools such as HERMES and NetID have addressed longstanding challenges in metabolite annotation. By focusing on molecular formulas and optimizing peak connections through integer linear programming, these methods enhance the accuracy and coverage of metabolomics datasets. The KGMN approach further extends annotation capabilities by integrating knowledge-based reaction networks, similarity metrics, and peak correlation data.

Finally, the chapter concludes with a call for standardized reporting practices in metabolomics. Guidelines proposed by Alseekh et al. aim to improve data quality, reproducibility, and cross-study comparability, paving the way for comprehensive pan-metabolome databases.

Together, these methodologies represent a paradigm shift in how bacterial metabolism is studied. They offer powerful tools for probing microbial function, diagnosing infections, and discovering therapeutic compounds. As the field continues to evolve, the integration of these approaches will be essential for advancing our understanding of microbial life and its impact on human health and the environment.

3.2 Machine-Generated Summaries

Disclaimer: The summaries in this chapter were generated from Springer Nature publications using abstractive AI auto-summarization: An abstraction-based summarizer creates new text based on deep learning. New phrases are created to summarize the content. As the constituted sentences are machine selected, they may not fully reflect the body of the work, so we strongly advise that the original content is read and cited. The auto-generated summaries were curated by the editor to meet Springer Nature publication standards.

Predictive Metabolomic Profiling of Microbial Communities Using Amplicon or Metagenomic Sequences [105]

This is a machine-generated summary of:

Mallick, Himel; Franzosa, Eric A.; McIver, Lauren J.; Banerjee, Soumya; Sirota-Madi, Alexandra; Kostic, Aleksandar D.; Clish, Clary B.; Vlamakis, Hera; Xavier, Ramnik J.; Huttenhower, Curtis Predictive Metabolomic Profiling of Microbial Communities Using Amplicon or Metagenomic Sequences [105]

Published in: Nature Communications (2019)

Link to original: https://doi.org/10.1038/s41467-019-10927-1

If you want to cite the papers, please refer to the original.

For technical reasons we could not place the page where the original quote is coming from.

Research Highlights

Advances in high-throughput metabolomics have significantly enhanced our ability to analyze small-molecule metabolites within microbial communities, providing insights into disease biomarkers and pathogenesis. These technologies have been pivotal in understanding the metabolic interactions within the human microbiome, particularly with bioactive metabolites like short-chain fatty acids and bile acids, which are linked to diseases such as inflammatory bowel disease (IBD) and colorectal cancer. Despite these advancements, a major challenge remains in predicting metabolomic activities from metagenomic data due to the incomplete repertoire of gene–metabolite reactions. Now, a new study by Himel Mallick and colleagues introduces MelonnPan, a computational framework designed to predict community metabolomes from microbial profiles, addressing this knowledge gap.

MelonnPan employs machine learning techniques to infer metabolite profiles from metagenomic or amplicon sequencing data without requiring detailed functional annotations. The framework uses elastic net regularization to identify predictive microbial features for each metabolite, allowing researchers to estimate composite metabolomes from taxonomic or functional profiles. The study applied MelonnPan to gut metagenome datasets from Crohn's disease patients, demonstrating high concordance between predicted and observed metabolic trends for various metabolites. The model was validated using independent datasets, showing robust prediction accuracy across different microbial environments, including coral-associated and murine gut communities. This approach enables researchers to infer metabolic profiles where only metagenomic data is available, offering a cost-effective alternative to direct metabolomic profiling.

The development of MelonnPan represents a significant advancement in linking microbial community profiles with their metabolic outputs. Future study directions include refining predictive accuracy by integrating additional microbial measurements such as metatranscriptomics and exploring more sophisticated machine learning models. While MelonnPan does not replace direct metabolomic profiling, it provides a valuable tool for hypothesis generation and experimental design in microbiome research. As reference databases expand and training datasets grow, the predictive capabilities of MelonnPan are expected to improve further, facilitating large-scale multi-omic analyses and enhancing our understanding of microbiome dynamics across various environments.

Genome-Resolved Metagenome and Metatranscriptome Analyses of Thermophilic Composting Reveal Key Bacterial Players and Their Metabolic Interactions [106]

This is a machine-generated summary of:

Braga, Lucas Palma Perez; Pereira, Roberta Verciano; Martins, Layla Farage; Moura, Livia Maria Silva; Sanchez, Fabio Beltrame; Patané, José Salvatore Leister; da Silva, Aline Maria; Setubal, João Carlos Genome-Resolved Metagenome and Metatranscriptome Analyses of Thermophilic Composting Reveal Key Bacterial Players and Their Metabolic Interactions [106]

Published in: BMC Genomics (2021)

Link to original: https://doi.org/10.1186/s12864-021-07957-9

If you want to cite the papers, please refer to the original.
For technical reasons we could not place the page where the original quote is coming from.

Research Highlights
Thermophilic composting is a process driven by diverse microbial communities capable of thriving in high-temperature environments. These microbes exhibit remarkable metabolic flexibility, efficiently breaking down complex organic matter such as lignocellulosic biomass. Despite the potential of composting microbiomes for biotechnological applications, understanding their functional diversity and metabolic interactions remains limited. Now, a new study by Lucas Palma Perez Braga and colleagues employs metagenome-assembled genomes (MAGs) to explore the microbial dynamics and metabolic interactions in a thermophilic composting process.

The study utilized time series samples from a composting facility at the São Paulo Zoo Park, employing shotgun metagenomic sequencing to recover MAGs from two composting cells. The researchers identified 60 MAGs, with 24 potentially representing novel bacterial species. These MAGs were taxonomically diverse, spanning six different phyla, and were analyzed for their functional capabilities using genome-scale metabolic models. The team focused on key processes such as lignocellulose degradation, denitrification, sulfur metabolism, hydrogen metabolism, and secondary metabolite production. By mapping metagenome reads to MAGs and analyzing metatranscriptomic data, the study revealed active transcriptional profiles of these microbes throughout the composting process. The researchers also identified syntrophic interactions among microbial populations, highlighting the exchange of metabolites like hydrogen and succinate.

The findings provide a comprehensive framework for understanding the temporal dynamics of microbial processes in thermophilic composting. The study suggests that specific bacterial populations play crucial roles at different stages of composting, with potential implications for enhancing biomass degradation efficiency. Future studies could focus on experimentally verifying the predicted metabolic interactions and exploring the application of these microbial communities as inoculants for industrial-scale biomass degradation. This research advances our understanding of composting microbiomes and sets the stage for engineering efficient thermophilic microbiomes for biotechnological applications.

Legionella *and* Coxiella *Effectors: Strength in Diversity and Activity [107]*

This is a machine-generated summary of:

Qiu, Jiazhang; Luo, Zhao-Qing *Legionella* and *Coxiella* Effectors: Strength in Diversity and Activity [107]

Published in: Nature Reviews Microbiology (2017)

Link to original: https://doi.org/10.1038/nrmicro.2017.67

If you want to cite the papers, please refer to the original.

For technical reasons we could not place the page where the original quote is coming from.

Research Highlights

Legionella pneumophila and *Coxiella burnetii* are intracellular pathogens responsible for Legionnaires' disease and Q fever, respectively. These bacteria reside and replicate within host cells in specialized compartments known as replication vacuoles. *L. pneumophila* is commonly found in freshwater environments and can cause severe pneumonia when inhaled by immunocompromised individuals. *C. burnetii* primarily infects humans through aerosols from livestock. Both pathogens utilize the Dot/Icm type IV secretion system to manipulate host cellular processes, facilitating their survival and replication within host cells. Now, a new study by Jiazhang Qiu and colleagues explores the functional characterization of Dot/Icm effectors from these pathogens, highlighting their roles in subverting host processes such as vesicle trafficking, lipid metabolism, and gene expression.

The study employs a combination of genetic, biochemical, and cell biology approaches to investigate how *L. pneumophila* and *C. burnetii* effectors manipulate host cell functions. The researchers identify numerous effectors that target key cellular pathways, including vesicle trafficking and ubiquitin signaling. For instance, *L. pneumophila* effectors like SidM and AnkX modulate the activity of small GTPases involved in vesicle transport, while others interfere with the host's ubiquitin network to alter protein stability and function. In contrast, *C. burnetii* effectors exhibit less redundancy but are crucial for optimal bacterial growth within host cells. The study also highlights the importance of lipid metabolism manipulation by these pathogens to maintain their replication vacuoles.

The advances made in understanding the roles of Dot/Icm effectors provide insights into bacterial pathogenesis and host–pathogen interactions. Future studies could focus on elucidating the biochemical activities of these effectors and their potential therapeutic applications. The research underscores the complexity of effector functions and suggests that further exploration could reveal novel mechanisms of bacterial virulence and host cell regulation. Additionally, understanding how these effectors contribute to nutrient acquisition by intravacuolar bacteria may offer new avenues for combating infections caused by *L. pneumophila* and *C. burnetii*.

A New HPLC-PAD/HPLC-ESI-MS Method for the Analysis of Phytoestrogens Produced by Bacterial Metabolism [108]

This is a machine-generated summary of:

Gaya, Pilar; Arqués, Juan Luis; Medina, Margarita; Álvarez, Inmaculada; Landete, José Maria A New HPLC-PAD/HPLC-ESI-MS Method for the Analysis of Phytoestrogens Produced by Bacterial Metabolism [108]

Published in: Food Analytical Methods (2015)

Link to original: https://doi.org/10.1007/s12161-015-0226-3

If you want to cite the papers, please refer to the original.

For technical reasons we could not place the page where the original quote is coming from.

Research Highlights

Phytoestrogens, plant-derived polyphenols with structural similarities to human estrogens, are found in various foods such as soybeans, flaxseeds, fruits, and vegetables. These compounds can exhibit both estrogenic and antiestrogenic effects and are metabolized by intestinal bacteria into more bioavailable forms like equol, urolithins, and enterolignans. These metabolites have enhanced estrogenic/antiestrogenic activity and antioxidant properties, contributing to protective effects against chronic diseases. Despite the known transformations of isoflavones by gut bacteria, the simultaneous analysis of isoflavones, lignans, and ellagitannins has been challenging. Now a new study by Pilar Gaya and colleagues presents a novel chromatographic method using HPLC-PAD to detect the transformation of these phytoestrogens by microorganisms.

The study developed an HPLC-PAD method to analyze the metabolism of isoflavones, lignans, and ellagitannins by intestinal bacteria. Researchers prepared standardized extracts from soybeans and flaxseeds and incubated them with human fecal samples under anaerobic conditions. The phytoestrogens were extracted using a combination of diethyl ether and ethyl acetate solvents before analysis. The HPLC-PAD system allowed for the simultaneous detection of various phytoestrogen metabolites such as equol from isoflavones, urolithin D from ellagitannins, and enterolactone from lignans. The method was validated for sensitivity and accuracy through comparison with known standards using HPLC-ESI/MS.

The study advances our understanding of phytoestrogen metabolism by providing an accessible method for detecting these transformations in complex environments like the gut microbiota. This approach could be applied in vivo to further explore the health implications of phytoestrogen consumption. Future studies could focus on refining this method for broader applications in dietary studies or clinical settings to assess individual variations in phytoestrogen metabolism. The findings underscore the importance of gut microbiota in modulating the bioactivity of dietary compounds and open avenues for personalized nutrition strategies based on microbial composition.

Real-Time Monitoring of Bacterial Biofilms Metabolic Activity by a Redox-Reactive Nanosensors Array [109]

This is a machine-generated summary of:

Yeor-Davidi, Ella; Zverzhinetsky, Marina; Krivitsky, Vadim; Patolsky, Fernando Real-Time Monitoring of Bacterial Biofilms Metabolic Activity by a Redox-Reactive Nanosensors Array [109]

Published in: Journal of Nanobiotechnology (2020)

Link to original: https://doi.org/10.1186/s12951-020-00637-y

If you want to cite the papers, please refer to the original.

For technical reasons we could not place the page where the original quote is coming from.

Research Highlights

Nanostructures like nanowires are pivotal in advancing electronic applications due to their unique electron-transport properties and high surface-to-volume ratio, making them ideal for field-effect-transistor (FET) devices. These properties enable nanowires to detect single molecules, offering potential for ultra-sensitive, selective, and label-free real-time sensors. However, practical application of nanowire-based FET devices in sensing is hindered by detection limitations such as Debye screening in high ionic strength conditions. Now, a new study by Ella Yeor-Davidi and colleagues addresses these challenges by developing a novel platform that monitors the metabolic activity of bacterial biofilms through glucose consumption using redox-reactive silicon nanowire FETs.

The study employs a unique surface modification of SiNW-FET devices with a redox system composed of 9,10-dihydroxyanthracene/9,10-anthraquinone, enabling real-time monitoring of glucose metabolism in high-ionic-strength solutions without preprocessing samples. The researchers utilized enzymatic reactions to convert metabolites into hydrogen peroxide, which interacts with the redox system to alter the conductivity of the FET device. This approach allows for continuous monitoring of biofilm processes in their natural environment. The study demonstrated the platform's capability by tracking glucose consumption in *Bacillus subtilis* biofilms and comparing it with conventional optical methods. The results showed that the SiNW-FET device could detect metabolic activity before changes were observable through optical means.

The advances made by this study highlight the potential of SiNW-FET arrays as promising tools for real-time, label-free monitoring of bacterial biofilm metabolism. Future studies could explore integrating this technology into medical implants to monitor bacterial activity and prevent infections. Additionally, further examination of biofilms under antibiotic treatment using SiNW-FET devices could provide insights into bioprocesses within biofilms and aid in developing effective treatments for biofilm-related infections. This novel sensing platform offers significant advantages over existing methods due to its sensitivity, selectivity, and noninvasive nature, paving the way for new applications in medical diagnostics and research on bacterial metabolism.

Metabolic Preference Assay for Rapid Diagnosis of Bloodstream Infections [110]

This is a machine-generated summary of:

Rydzak, Thomas; Groves, Ryan A.; Zhang, Ruichuan; Aburashed, Raied; Pushpker, Rajnigandha; Mapar, Maryam; Lewis, Ian A. Metabolic Preference Assay for Rapid Diagnosis of Bloodstream Infections [110]

Published in: Nature Communications (2022)
Link to original: https://doi.org/10.1038/s41467-022-30048-6

If you want to cite the papers, please refer to the original.

For technical reasons we could not place the page where the original quote is coming from.

Research Highlights

The diagnostic tools used to identify pathogens and measure antimicrobial susceptibility are crucial in managing infectious diseases, particularly bloodstream infections (BSIs), where rapid diagnosis is vital. Delays in appropriate antimicrobial treatment can significantly increase mortality rates, especially when BSIs progress to septic shock. Current diagnostic methods often take several days, leading to incorrect antimicrobial prescriptions for a significant percentage of patients. Now, a new study by Thomas Rydzak and colleagues introduces a metabolomics-based diagnostic strategy that could significantly reduce the time required for pathogen identification and antimicrobial susceptibility testing (AST).

The study presents a novel approach called the metabolic preference assay (MPA), which utilizes patterns of consumed and excreted metabolites from ex vivo microbial cultures to identify pathogens and assess their antimicrobial susceptibility. The researchers identified seven metabolites capable of differentiating between common BSI pathogens such as *Candida albicans*, *Pseudomonas aeruginosa*, and *Streptococcus pneumoniae*. The metabolic inhibition assay (MIA) was then used to measure changes in metabolite levels in the presence of antimicrobials, providing rapid AST results. The study involved analyzing metabolic boundary fluxes under controlled conditions using ultra-high-performance liquid chromatography-mass spectrometry (UHPLC-MS). The MPA/MIA workflow demonstrated a significant reduction in diagnostic timelines compared to traditional methods.

The MPA/MIA diagnostic workflow offers several advantages for clinical diagnostics, including integration of identification and AST on a single platform with minimal sample handling. It allows direct analysis from blood culture bottles without the need for subculturing isolates. Future study could expand this approach to cover a broader range of pathogens and optimize the workflow for high-throughput clinical settings. The transition to clinical-grade mass spectrometers could further enhance its applicability in real-world laboratories. This metabolomics-based strategy represents a promising advancement in rapid diagnostics for infectious diseases, potentially improving patient outcomes and promoting antibiotic stewardship by enabling more precise therapies.

Microbial Metabolomics in Open Microscale Platforms [111]

This is a machine-generated summary of:

Barkal, Layla J.; Theberge, Ashleigh B.; Guo, Chun-Jun; Spraker, Joe; Rappert, Lucas; Berthier, Jean; Brakke, Kenneth A.; Wang, Clay C. C.; Beebe, David J.; Keller, Nancy P.; Berthier, Erwin Microbial Metabolomics in Open Microscale Platforms [111]

Published in: Nature Communications (2016)

Link to original: https://doi.org/10.1038/ncomms10610

Research Highlights

Microbial secondary metabolism is a rich source of bioactive compounds with significant implications for health and disease management across various domains. Filamentous fungi, in particular, are known for producing secondary metabolites that serve as virulence factors and therapeutic agents. However, the exploration of these metabolites is hindered by the complexity of their production, which is often triggered by specific environmental cues. Traditional methods for culturing fungi and isolating metabolites are labor-intensive and inefficient, limiting the discovery of novel compounds. Now, a new study by Layla J. Barkal and colleagues presents an innovative microscale platform that addresses these challenges by enabling efficient exploration of fungal secondary metabolism under diverse microenvironmental conditions.

The study introduces a microscale metabolomics platform that utilizes open microfluidic channels to culture microorganisms and extract metabolites passively.

This system employs micro-agar pads or liquid wells within teardrop-shaped channels to maintain microbial cultures while directing solvent flow over the culture area to form stable biphasic interfaces. The design leverages surface tension to ensure precise solvent retrieval without contamination from aqueous media, facilitating reproducible metabolite extraction suitable for LC-MS analysis. The platform's open nature allows for air exposure, which is crucial for certain fungal processes like sporulation. By using this device, the researchers conducted multidimensional experiments with *Aspergillus nidulans* as a model system, demonstrating the platform's capability to efficiently extract metabolites under varied conditions and even enabling coculture studies with bacteria.

The micrometabolomics platform represents a significant advancement in microbial metabolomics research by offering a compact, efficient alternative to traditional methods. Its ability to handle small volumes and diverse solvents opens new avenues for exploring previously inaccessible segments of the metabolome. Future study directions could include expanding the platform's use to more complex cocultures or rare biological samples, potentially uncovering novel secondary metabolites relevant to disease progression or therapeutic development. The integration of this technology into routine laboratory workflows could accelerate the discovery of bioactive compounds and enhance our understanding of microbial interactions in natural environments.

HERMES: A Molecular-Formula-Oriented Method to Target the Metabolome [112]

This is a machine-generated summary of:

Giné, Roger; Capellades, Jordi; Badia, Josep M.; Vughs, Dennis; Schwaiger-Haber, Michaela; Alexandrov, Theodore; Vinaixa, Maria; Brunner, Andrea M.; Patti, Gary J.; Yanes, Oscar HERMES: A Molecular-Formula-Oriented Method to Target the Metabolome [112]

Published in: Nature Methods (2021)

Link to original: https://doi.org/10.1038/s41592-021-01307-z

If you want to cite the papers, please refer to the original.

For technical reasons we could not place the page where the original quote is coming from.

Research Highlights

Metabolomic experiments using liquid chromatography–mass spectrometry (LC-MS) often generate vast amounts of data, which include numerous redundant ions and artifacts that complicate the identification of biologically relevant metabolites. Traditional untargeted metabolomics approaches face challenges due to heterogeneous chromatographic peak shapes and stochastic selection of precursor ions, leading to low reproducibility and poor spectral quality. Now, a new study by Roger Giné and colleagues introduces HERMES, a novel method and computational tool designed to enhance the selectivity and sensitivity of metabolite profiling by focusing on sample-specific signals.

HERMES employs an inverse approach to conventional workflows by directly interrogating raw LC-MS data using a comprehensive list of unique molecular formulas tailored to the sample's nature. These formulas are derived from extensive databases or specific metabolic pathways. The method generates multiple ionic formulas for each molecular formula, which are then matched against millions of data points in an LC-MS experiment. HERMES predicts isotopic patterns for these ionic formulas, annotates isotope signals, and calculates isotopic fidelity. It identifies clusters of data points, termed Scans of Interest (SOIs), that match ionic formulas within a short time frame, independent of peak shape assumptions. The workflow includes blank subtraction using neural networks, adduct grouping based on elution profiles, and annotation with low-energy data. Users can prioritize SOIs for targeted acquisition based on criteria like adduct type and isotopic fidelity.

The introduction of HERMES represents a significant advancement in metabolomics by addressing the limitations of traditional peak detection methods. It improves metabolite identification by focusing on biologically relevant compounds and providing cleaner spectra through continuous targeted acquisition without dynamic exclusion. Future study could optimize ion injection times and collision energies for inclusion list entries to enhance spectral quality further. Additionally, expanding public databases will improve the annotation of in-source fragments not currently covered. HERMES offers flexibility by allowing users to incorporate new molecular formulas beyond existing databases, potentially broadening its application across various fields in biomedical and environmental research.

Exploiting Targeted and Untargeted Approaches for the Analysis of Bacterial Metabolites Under Altered Growth Conditions [113]

This is a machine-generated summary of:

Zanella, Delphine; Liden, Tiffany; York, Jamie; Franchina, Flavio A.; Focant, Jean-François; Schug, Kevin A. Exploiting Targeted and Untargeted Approaches for the Analysis of Bacterial Metabolites Under Altered Growth Conditions [113]

Published in: Analytical and Bioanalytical Chemistry (2021)
Link to original: https://doi.org/10.1007/s00216-021-03505-2

If you want to cite the papers, please refer to the original.

For technical reasons we could not place the page where the original quote is coming from.

Research Highlights

The identification of microorganisms is crucial in clinical, biological, and food applications, such as diagnosing chronic infectious diseases and ensuring food safety. Accurate diagnosis of infectious diseases can reduce their incidence and improve human health. Pathogenic microorganisms have developed adaptive defense mechanisms, known as stress responses, to survive in changing environments, which complicates their identification. Now a new study by Delphine Zanella and colleagues explores the use of metabolomics to differentiate bacterial species under various stress conditions.

The study employed both targeted and untargeted metabolomics approaches using high-performance liquid chromatography-mass spectrometry (HPLC-MS) and ultra-high-performance liquid chromatography-mass spectrometry (UHPLC-MS) platforms. The researchers investigated the metabolism of three bacterial species: *Staphylococcus aureus, Bacillus cereus*, and *Pseudomonas aeruginosa*. These bacteria were grown under different stress conditions, including acidic, basic, oxidative, ethanolic, and saline environments. The targeted analysis focused on primary metabolites using HPLC-QqQ MS to identify species-specific metabolic profiles under normal growth conditions. In contrast, the untargeted UHPLC-QToF MS approach allowed for a comprehensive analysis of the metabolome under stress conditions. Statistical analyses were used to reduce data complexity and identify significant metabolites that could differentiate between the bacterial species.

The combination of targeted and untargeted metabolomics proved valuable for differentiating bacterial species under normal and stress conditions. While targeted analysis effectively identified primary metabolites for distinguishing bacteria in normal conditions, it was less effective under stress. The untargeted approach provided a more comprehensive differentiation of bacterial species across various stress responses. Future studies could focus on validating the identification of species-specific metabolites with standards and correlating them with metabolic pathways. This could lead to developing a targeted method for faster profiling of microorganisms in diverse applications such as medical diagnostics and food safety.

Metabolite Discovery Through Global Annotation of Untargeted Metabolomics Data [114]

This is a machine-generated summary of:

Chen, Li; Lu, Wenyun; Wang, Lin; Xing, Xi; Chen, Ziyang; Teng, Xin; Zeng, Xianfeng; Muscarella, Antonio D.; Shen, Yihui; Cowan, Alexis; McReynolds, Melanie R.; Kennedy, Brandon J.; Lato, Ashley M.; Campagna, Shawn R.; Singh, Mona; Rabinowitz, Joshua D. Metabolite Discovery Through Global Annotation of Untargeted Metabolomics Data [114]

Published in: Nature Methods (2021)
Link to original: https://doi.org/10.1038/s41592-021-01303-3

If you want to cite the papers, please refer to the original.

For technical reasons we could not place the page where the original quote is coming from.

Research Highlights

Metabolomics is a powerful tool that provides a comprehensive snapshot of small molecule concentrations within biological systems, reflecting the combined effects of genetics and environment on metabolism. Despite advancements in liquid chromatography–high-resolution mass spectrometry (LC-MS) techniques, a significant challenge remains in annotating the vast number of unknown peaks detected in metabolomics studies. Now, a new study by Li Chen and colleagues introduces an innovative approach using integer linear programming to enhance the annotation accuracy of metabolomics data.

The study presents NetID, a stand-alone algorithm designed to optimize the network of mass spectrometry peak connections by leveraging mass differences and spectral similarities. The method differentiates between biochemical connections and those arising from mass spectrometry phenomena, incorporating known metabolite data and retention times. The researchers applied this approach to untargeted metabolomics data from baker's yeast and mouse liver, using both positive and negative ionization modes. The global optimization step in NetID enforces a single formula assignment for each ion peak, significantly improving annotation accuracy as validated by a target-decoy strategy. This process resulted in probable formulae for several hundred potential metabolites that were previously unannotated and confirmed the identities of five previously unrecognized metabolites.

The introduction of integer linear programming into metabolomics data annotation represents a significant advancement in the field. By considering all peak annotations simultaneously, NetID maximizes the use of available information, providing more accurate and comprehensive annotations than traditional methods. Future study could focus on expanding the list of feasible mass spectrometry adducts to account for uncommon adducts that may explain many unannotated peaks. Additionally, integrating concepts from other computational pipelines could further enhance annotation accuracy. As experimental validation continues to support these computational predictions, this approach holds promise for identifying most unknown metabolites over the coming decade, offering a robust blueprint of the metabolome.

Prediction and Quantification of Bioactive Microbiota Metabolites in the Mouse Gut [115]

This is a machine-generated summary of:

Sridharan, Gautham V.; Choi, Kyungoh; Klemashevich, Cory; Wu, Charmian; Prabakaran, Darshan; Pan, Long Bin; Steinmeyer, Shelby; Mueller, Carrie; Yousofshahi, Mona; Alaniz, Robert C.; Lee, Kyongbum; Jayaraman, Arul Prediction and Quantification of Bioactive Microbiota Metabolites in the Mouse Gut [115]

Published in: Nature Communications (2014)

Link to original: https://doi.org/10.1038/ncomms6492

If you want to cite the papers, please refer to the original.

For technical reasons we could not place the page where the original quote is coming from.

Research Highlights

The human gastrointestinal tract is home to a diverse array of microorganisms collectively known as the microbiota. These microorganisms play a crucial role in maintaining gut health, and disruptions in their composition have been linked to various diseases, including obesity and diabetes. The metabolites produced by the microbiota are increasingly recognized as important modulators of host physiology. However, identifying these bioactive metabolites remains challenging due to the complexity of the gut environment and the difficulty in distinguishing between host- and microbiota-derived metabolites. Now, a new study by Gautham V. Sridharan and colleagues presents a novel approach that integrates computational predictions with targeted metabolomics to identify bioactive metabolites derived from aromatic amino acids (AAAs) in the mouse gut.

The study employs a combined host–microbiota reaction network model to predict potential metabolic pathways involving AAAs, such as tryptophan, phenylalanine, and tyrosine. Using a probabilistic pathway construction algorithm, the researchers identified metabolites that are likely products of microbiota metabolism. To validate these predictions, they conducted targeted metabolomics experiments using mass spectrometry on cecum contents from conventionally raised specific pathogen-free (SPF) mice and germ-free (GF) mice. The analysis revealed several metabolites that were significantly reduced or absent in GF mice, indicating their microbial origin. Additionally, the study explored the bioactivity of these metabolites by assessing their ability to activate the aryl hydrocarbon receptor (AhR), a key regulator of immune function.

This research advances our understanding of gut microbiota metabolism by providing a framework for identifying bioactive microbial metabolites and their potential roles in host physiology. The integration of computational predictions with experimental validation offers a powerful tool for exploring complex metabolic interactions in the gut. Future studies could expand this approach to investigate other source metabolites or physiological conditions, potentially uncovering new biomarkers for diseases associated with microbiota alterations. Incorporating genomic data into pathway analysis could further refine predictions by accounting for gene abundance and expression levels, enhancing our ability to link specific microbial functions to health outcomes.

Metabolite Annotation from Knowns to Unknowns Through Knowledge-Guided Multi-Layer Metabolic Networking [116]

This is a machine-generated summary of:

Zhou, Zhiwei; Luo, Mingdu; Zhang, Haosong; Yin, Yandong; Cai, Yuping; Zhu, Zheng-Jiang Metabolite Annotation from Knowns to Unknowns Through Knowledge-Guided Multi-Layer Metabolic Networking [116]

Published in: Nature Communications (2022)

Link to original: https://doi.org/10.1038/s41467-022-34537-6

If you want to cite the papers, please refer to the original.

For technical reasons we could not place the page where the original quote is coming from.

Research Highlights

The metabolome encompasses the complete collection of small molecules within living organisms, including both known and unknown metabolites derived from various sources such as microbiota, plants, foods, and xenobiotics. Liquid chromatography–mass spectrometry (LC–MS) based untargeted metabolomics can measure thousands of metabolic features from biological samples, but the identification of these metabolites remains a significant challenge. Traditional methods rely on spectral libraries for known metabolites, while unknown metabolites require additional experiments or in silico tools. Now, a new study by Zhiwei Zhou and colleagues introduces a knowledge-guided multilayer network (KGMN) approach to enhance global metabolite annotation from knowns to unknowns in untargeted metabolomics.

The KGMN approach integrates three layers of networks: a knowledge-based metabolic reaction network, a knowledge-guided similarity network, and a global peak correlation network. Initially, seed metabolites are annotated by matching their properties to standard libraries and mapped into the metabolic reaction network to identify reaction-paired neighbor metabolites. This network includes both known and in silico-generated unknown metabolites linked through reactions. The knowledge-guided similarity network then uses these connections to match experimental data with predicted values for m/z and retention times (RT), while also considering MS/MS spectral similarity. Finally, the global peak correlation network identifies different ion forms in LC–MS data by utilizing chromatography co-elution correlations. This multilayer strategy significantly improves the accuracy of metabolite identification and extends annotation coverage.

The KGMN approach represents a significant advancement in metabolite annotation by integrating multiple data-driven and knowledge-based networks. It allows for efficient annotation of unknown metabolites and enhances the discovery of recurrent unknowns in biological samples. Future studies could focus on incorporating more orthogonal properties such as collision cross-section to further improve the accuracy of chemical structure predictions for unknowns. Additionally, expanding the knowledge base with reactions from nonhuman species could broaden the applicability of KGMN to diverse biological contexts. The integration of repository mining with KGMN also offers potential for validating recurrent unknowns and gaining deeper insights into their biological functions.

Mass Spectrometry-Based Metabolomics: A Guide for Annotation, Quantification and Best Reporting Practices [117]

This is a machine-generated summary of:

Alseekh, Saleh; Aharoni, Asaph; Brotman, Yariv; Contrepois, Kévin; D'Auria, John; Ewald, Jan; C. Ewald, Jennifer; Fraser, Paul D.; Giavalisco, Patrick; Hall, Robert D.; Heinemann, Matthias; Link, Hannes; Luo, Jie; Neumann, Steffen; Nielsen, Jens; Perez de Souza, Leonardo; Saito, Kazuki; Sauer, Uwe; Schroeder, Frank C.; Schuster, Stefan; Siuzdak, Gary; Skirycz, Aleksandra; Sumner, Lloyd W.; Snyder, Michael P.; Tang, Huiru; Tohge, Takayuki; Wang, Yulan; Wen, Weiwei; Wu, Si; Xu, Guowang; Zamboni, Nicola; Fernie, Alisdair R. Mass Spectrometry-Based Metabolomics: A Guide for Annotation, Quantification and Best Reporting Practices [117]

Published in: Nature Methods (2021)

Link to original: https://doi.org/10.1038/s41592-021-01197-1

If you want to cite the papers, please refer to the original.
For technical reasons we could not place the page where the original quote is coming from.

Research Highlights
Metabolomics, the large-scale study of metabolites within cells, has become a vital tool in systems biology, aiding research across plant, microbial, and mammalian domains. Despite its maturity and widespread application, the field faces significant challenges in achieving comprehensive metabolome coverage due to the vast diversity and rapid turnover of metabolites. Now, a new study by Saleh Alseekh and colleagues addresses these challenges by proposing simplified reporting guidelines to enhance the acquisition and interpretation of metabolomics data.

The study outlines a streamlined workflow focusing on chromatography coupled with mass spectrometry (MS), which is prevalent in metabolomics research. The authors emphasize the importance of rigorous sampling, extraction, and storage protocols to ensure data quality. They propose guidelines for reporting processed data rather than raw chromatograms, aiming to improve the robustness of metabolite identification and quantification. The study also highlights the necessity of recombination and recovery experiments to validate metabolite annotations and suggests stricter nomenclature standards to reduce ambiguity in reporting. By focusing on these aspects, Alseekh et al. aim to address common pitfalls in metabolomics studies, such as ion suppression and peak misidentification.

The advances proposed by Alseekh and colleagues are expected to facilitate better cross-laboratory comparisons of metabolomics datasets, akin to practices in transcriptomics that have yielded significant biological insights. Future study directions include expanding these guidelines to encompass computational analysis tools and integrating them with existing repository standards like MetaboLights. The authors advocate for broader adoption of their recommendations to improve data transparency and comparability across different species and research contexts. This approach could lead to the development of comprehensive pan-metabolome databases, enhancing our understanding of metabolic processes across various life forms.

3.3 Conclusion

The study of bacterial metabolism has entered a transformative era, driven by the convergence of high-resolution analytical techniques, computational modeling, and integrative multi-omics approaches. Chapter 3 has explored a wide array of methodologies that collectively redefine how researchers investigate microbial metabolic processes. These tools not only enhance our capacity to characterize bacterial physiology but also open new avenues for diagnostics, therapeutics, and ecological engineering.

One of the central themes of this chapter is the increasing reliance on predictive and computational frameworks to infer metabolic activity from genomic data. The MelonnPan model exemplifies this shift, enabling researchers to estimate metabolite profiles from microbial community sequencing without direct metabolomic measurements. This approach is particularly valuable in large-scale microbiome studies, where resource constraints often limit the feasibility of comprehensive metabolite profiling. By leveraging machine learning and elastic net regularization, MelonnPan bridges the gap between taxonomic composition and functional output, offering a scalable solution for hypothesis generation and experimental design.

Complementing these predictive tools are genome-resolved metagenomic and metatranscriptomic analyses, which provide detailed insights into microbial community dynamics and metabolic interactions. The study of thermophilic composting microbiomes illustrates how metagenome-assembled genomes (MAGs) and transcriptional profiling can reveal syntrophic relationships and functional specialization among bacterial populations. These findings have practical implications for biotechnological applications, such as optimizing biomass degradation and engineering microbial consortia for industrial processes.

The chapter also highlights the importance of studying intracellular pathogens, *such as Legionella pneumophila* and *Coxiella burnetii*, whose metabolic strategies are linked to host cell manipulation. Investigations into their Dot/Icm effector systems reveal how these bacteria hijack host lipid metabolism and vesicle trafficking to sustain replication. Understanding these interactions not only informs pathogenesis but also identifies potential metabolic vulnerabilities that could be exploited for therapeutic intervention.

Analytical chemistry continues to play a pivotal role in bacterial metabolomics. The development of HPLC-PAD and HPLC-ESI-MS methods for detecting phytoestrogen transformations by gut bacteria underscores the value of precise, sensitive assays in uncovering microbial contributions to host health. These methods facilitate the study of bioactive dietary compounds and their microbial metabolites, with implications for personalized nutrition and chronic disease prevention.

Technological innovation is further exemplified by the use of redox-reactive nanosensor arrays for real-time monitoring of bacterial biofilm metabolism. The SiNW-FET platform enables label-free detection of glucose consumption, offering a noninvasive and highly sensitive method for tracking metabolic activity in situ. This technology holds promise for medical diagnostics, particularly in monitoring implant-associated infections and evaluating biofilm responses to antimicrobial treatments.

Rapid diagnostics are another area where metabolomics is making a significant impact. The metabolic preference assay (MPA) and metabolic inhibition assay (MIA) offer integrated workflows for pathogen identification and antimicrobial susceptibility testing. By analyzing metabolite fluxes directly from blood cultures, these assays reduce diagnostic timelines and support more targeted antimicrobial therapy, addressing a critical need in clinical microbiology.

Microscale platforms have revolutionized the study of microbial secondary metabolism, particularly in filamentous fungi. Open microfluidic systems allow for controlled culturing and solvent extraction, enabling the discovery of novel bioactive compounds under diverse environmental conditions. These platforms are especially valuable for coculture experiments, where microbial interactions can trigger the production of previously inaccessible metabolites.

Computational tools such as HERMES, NetID, and KGMN have significantly advanced metabolite annotation, addressing the challenges posed by complex LC-MS datasets. By integrating molecular formula databases, spectral similarity networks, and peak correlation analyses, these tools improve the accuracy and coverage of metabolomics studies. Their flexibility and scalability make them indispensable for untargeted metabolomics and global metabolite discovery.

Finally, the chapter emphasizes the need for standardized reporting practices in metabolomics. Guidelines proposed by Alseekh et al. aim to enhance data transparency, reproducibility, and cross-study comparability. These standards are essential for building robust pan-metabolome databases and facilitating collaborative research across disciplines.

In conclusion, the methodologies presented in this chapter represent a comprehensive toolkit for studying bacterial metabolism in different contexts. They empower researchers to move beyond descriptive analyses toward predictive, mechanistic, and translational insights. As these tools continue to evolve and integrate, they will play a central role in advancing microbiology, improving human health, and addressing global challenges in biotechnology and infectious disease management.

Bibliography

1. Shreiner, A. B., Kao, J. Y. & Young, V. B. The gut microbiome in health and in disease. *Curr. Opin. Gastroenterol. 31*, 69–75 (2015).
2. Langille, M. G. et al. Predictive functional profiling of microbial communities using 16S rRNA marker gene sequences. *Nat. Biotechnol. 31*, 814–821 (2013).
3. Subramanian, A. et al. Gene set enrichment analysis: a knowledge-based approach for interpreting genome-wide expression profiles. *Proc. Natl Acad. Sci. USA 102*, 15545–15550 (2005).
4. Zwiener, I., Frisch, B. & Binder, H. Transforming RNA-Seq data to improve the performance of prognostic gene signatures. *PLoS ONE 9*, e85150 (2014).
5. Zhou, X. & Stephens, M. Efficient multivariate linear mixed model algorithms for genome-wide association studies. *Nat. Methods 11*, 407–409 (2014).
6. Noecker, C. et al. Metabolic Model-based integration of microbiome taxonomic and metabolomic profiles elucidates mechanistic links between ecological and metabolic variation. *mSystems* https://doi.org/10.1128/mSystems.00013-15 (2016).
7. Theriot, C. M. et al. Antibiotic-induced shifts in the mouse gut microbiome and metabolome increase susceptibility to *Clostridium difficile* infection. *Nat. Commun. 5*, 3114 (2014).
8. Antunes LP, Martins LF, Pereira RV, Thomas AM, Barbosa D, Lemos LN, et al. Microbial community structure and dynamics in thermophilic composting viewed through metagenomics and metatranscriptomics. Sci Rep. 2016;6:ARTN 38915. https://doi.org/10.1038/srep38915.

9. Ma L, Zhao Y, Meng L, Wang X, Yi Y, Shan Y, et al. Isolation of Thermostable Lignocellulosic Bacteria From Chicken Manure Compost and a M42 Family Endocellulase Cloning From *Geobacillus thermodenitrificans* Y7. Front Microbiol. 2020;11:ARTN 281. https://doi.org/10.3389/fmicb.2020.00281.
10. Kolinko S, Wu YW, Tachea F, Denzel E, Hiras J, Gabriel R, et al. A bacterial pioneer produces cellulase complexes that persist through community succession. Nat Microbiol. 2018;3(1). https://doi.org/10.1038/s41564-017-0052-z.
11. Jimenez DJ, Dini-Andreote F, DeAngelis KM, Singer SW, Salles JF, van Elsas JD. Ecological insights into the dynamics of plant biomass-degrading microbial consortia. Trends Microbiol. 2017;25(10):788–96. https://doi.org/10.1016/j.tim.2017.05.012.
12. Palaniveloo K, Amran MA, Norhashim NA, Mohamad-Fauzi N, Peng-Hui F, Hui-Wen L, et al. Food waste composting and microbial community structure profiling. Processes. 2020;8(6). https://doi.org/10.3390/pr8060723.
13. Silva NM, de Oliveira A, Pegorin S, Giusti CE, Ferrari VB, Barbosa D, et al. Characterization of novel hydrocarbon-degrading *Gordonia paraffinivorans* and *Gordonia sihwensis* strains isolated from composting. PLoS One. 2019;14(4):16. https://doi.org/10.1371/journal.pone.0215396.
14. Lutz S, Thuerig B, Oberhaensli T, Mayerhofer J, Fuchs JG, Widmer F, et al. Harnessing the microbiomes of suppressive composts for plant protection: from metagenomes to beneficial microorganisms and reliable diagnostics. Front Microbiol. 2020;11. https://doi.org/10.3389/fmicb.2020.01810.
15. Weimann A, Mooren K, Frank J, Pope PB, Bremges A, McHardy AC. From genomes to phenotypes: traitar, the microbial trait analyzer. Msystems. 2016;1(6):ARTN e00101. https://doi.org/10.1128/mSystems.00101-16.
16. Uritskiy GV, DiRuggiero J, Taylor J. MetaWRAP-a flexible pipeline for genome-resolved metagenomic data analysis. Microbiome. 2018;6:ARTN 158. https://doi.org/10.1186/s40168-018-0541-1.
17. Quinlan, A. R. & Hall, I. M. BEDTools: a flexible suite of utilities for comparing genomic features. *Bioinforma. Oxf. Engl. 26*, 841–842 (2010).
18. Faust K, Raes J. CoNet app: inference of biological association networks using Cytoscape. F1000Research. 2016. https://doi.org/10.12688/f1000research.9050.2.
19. Shannon P, Markiel A, Ozier O, Baliga NS, Wang JT, Ramage D, et al. Cytoscape: a software environment for integrated models of biomolecular interaction networks. Genome Res. 2003;13(11):2498–504. https://doi.org/10.1101/gr.1239303.
20. Moffatt, J. H., Newton, P. & Newton, H. J. *Coxiella burnetii*: turning hostility into a home. *Cell. Microbiol. 17*, 621–631 (2015).
21. Weber, M. M. et al. Identification of *Coxiella burnetii* type IV secretion substrates required for intracellular replication and *Coxiella*-containing vacuole formation. *J. Bacteriol. 195*, 3914–3924 (2013).
22. Omsland, A. et al. Host cell-free growth of the Q fever bacterium *Coxiella burnetii. Proc. Natl Acad. Sci. USA 106*, 4430–4434 (2009). *This paper establishes, for the first time, a method for axenic culturing of C. burnetii , making later genetic manipulation of this pathogen possible.*
23. Chen, C. et al. Large-scale identification and translocation of type IV secretion substrates by *Coxiella burnetii. Proc. Natl Acad. Sci. USA 107*, 21755–21760 (2010). *In addition to identifying numerous effectors, this paper first expressed exogenous proteins in C. burnetii and demonstrated protein translocation by its Dot/Icm system.*
24. Martinez, E. et al. *Coxiella burnetii* effector CvpB modulates phosphoinositide metabolism for optimal vacuole development. *Proc. Natl Acad. Sci. USA 113*, E3260–E3269 (2016).
25. Hardiman, C. A., McDonough, J. A., Newton, H. J. & Roy, C. R. The role of Rab GTPases in the transport of vacuoles containing *Legionella pneumophila* and *Coxiella burnetii. Biochem. Soc. Trans. 40*, 1353–1359 (2012).

26. Nagai, H., Kagan, J. C., Zhu, X., Kahn, R. A. & Roy, C. R. A bacterial guanine nucleotide exchange factor activates ARF on *Legionella* phagosomes. *Science 295*, 679–682 (2002).
27. Ingmundson, A., Delprato, A., Lambright, D. G. & Roy, C. R. *Legionella pneumophila* proteins that regulate Rab1 membrane cycling. *Nature 450*, 365–369 (2007).
28. Arasaki, K., Toomre, D. K. & Roy, C. R. The *Legionella pneumophila* effector DrrA is sufficient to stimulate SNARE-dependent membrane fusion. *Cell Host Microbe 11*, 46–57 (2012).
29. Simon, S., Wagner, M. A., Rothmeier, E., Muller-Taubenberger, A. & Hilbi, H. Icm/Dot-dependent inhibition of phagocyte migration by *Legionella* is antagonized by a translocated Ran GTPase activator. *Cell. Microbiol. 16*, 977–992 (2014).
30. Rothmeier, E. et al. Activation of Ran GTPase by a *Legionella* effector promotes microtubule polymerization, pathogen vacuole motility and infection. *PLoS Pathog. 9*, e1003598 (2013).
31. Qiu, J. et al. Ubiquitination independent of E1 and E2 enzymes by bacterial effectors. *Nature 533*, 120–124 (2016). *This paper demonstrates, for the first time, that ubiquitylation can be catalysed by a single protein without the need for the E1 and E2 enzymes and ATP in a process in which ubiquitin is activated by ADP-ribosylation.*
32. Kotewicz, K. M. et al. A single *Legionella* effector catalyzes a multistep ubiquitination pathway to rearrange tubular endoplasmic reticulum for replication. *Cell Host Microbe 21*, 169–181 (2017). *Together with reference 80, this paper reveals the need of phosphodiesterase activity for the ubiquitylation reaction induced by the SidE family effectors; it also demonstrates that ubiquitylation of RTN4 by SidE family members leads to its association with the LCV.*
33. Qiu, J. et al. A unique deubiquitinase that deconjugates phosphoribosyl-linked protein ubiquitination. *Cell Res.* https://doi.org/10.1038/cr.2017.66 (2017).
34. Weber, S. S., Ragaz, C. & Hilbi, H. The inositol polyphosphate 5-phosphatase OCRL1 restricts intracellular growth of *Legionella*, localizes to the replicative vacuole and binds to the bacterial effector LpnE. *Cell. Microbiol. 11*, 442–460 (2009).
35. Banga, S. et al. *Legionella pneumophila* inhibits macrophage apoptosis by targeting pro-death members of the Bcl2 protein family. *Proc. Natl Acad. Sci. USA 104*, 5121–5126 (2007).
36. Hsu, F. et al. Structural basis for substrate recognition by a unique *Legionella* phosphoinositide phosphatase. *Proc. Natl Acad. Sci. USA 109*, 13567–13572 (2012). *This paper reveals a mechanism used by L. pneumophila to produce PtdIns4P on the LCV.*
37. Dong, N. et al. Modulation of membrane phosphoinositide dynamics by the phosphatidylinositide 4-kinase activity of the *Legionella* LepB effector. *Nat. Microbiol. 2*, 16236 (2016). *Together with reference 94, this paper demonstrates the coordination of two effectors for the production of PtdIns4P on the LCV.*
38. Losick, V. P. & Isberg, R. R. NF-κB translocation prevents host cell death after low-dose challenge by *Legionella pneumophila. J. Exp. Med. 203*, 2177–2189 (2006).
39. Rolando, M. et al. *Legionella pneumophila* effector RomA uniquely modifies host chromatin to repress gene expression and promote intracellular bacterial replication. *Cell Host Microbe 13*, 395–405 (2013).
40. Ge, J. et al. A *Legionella* type IV effector activates the NF-κB pathway by phosphorylating the IκB family of inhibitors. *Proc. Natl Acad. Sci. USA 106*, 13725–13730 (2009).
41. Losick, V. P., Haenssler, E., Moy, M. Y. & Isberg, R. R. LnaB: a *Legionella pneumophila* activator of NF-κB. *Cell. Microbiol. 12*, 1083–1097 (2010).
42. Weber, M. M. et al. Modulation of the host transcriptome by *Coxiella burnetii* nuclear effector Cbu1314. *Microbes Infect. 18*, 336–345 (2016).
43. Colonne, P. M. et al. Vasodilator-stimulated phosphoprotein activity is required for *Coxiella burnetii* growth in human macrophages. *PLoS Pathog. 12*, e1005915 (2016).
44. Lifshitz, Z. et al. Identification of novel *Coxiella burnetii* Icm/Dot effectors and genetic analysis of their involvement in modulating a mitogen-activated protein kinase pathway. *Infect. Immun. 82*, 3740–3752 (2014).
45. Vitale DC, Melilli B, Salomone S (2013) Isoflavones: estrogenic activity, biological effect and bioavailability. Eur J Drug Metab Pharmacokinet 38:15–25

46. Larrosa M, González-Sarrías A, García-Conesa MT, Tomás-Barberán FA, Espín JC (2006) Urolithins, ellagic acid-rerived metabolites produced by human colonic microflora, exhibit estrogenic and antiestrogenic activities. J Agric Food Chem 54:1611–1620
47. Zubik L, Meydani M (2003) Bioavailability of soybean isoflavones from aglycone and glucoside forms in American women. Am J Clin Nutr 77:1459–1465
48. Decroos K, Vanhemmens S, Cattoir S, Boon N, Verstraete W (2005) Isolation and characterisation of an equol-producing mixed microbial culture from a human faecal sample and its activity under gastrointestinal conditions. Arch Microbiol 183:45–55
49. Borges G, Roowi S, Rouanet J-M, Duthie GG, Lean MEJ, Crozier A (2007) The bioavailability of raspberry anthocyanins and ellagitannins in rats. Mol Nutr Food Res 51:714–725
50. Landete JM (2011) Ellagitannins, ellagic acid and their derived metabolites: a review about source, metabolism, functions and health. Food Res Int 44:1150–11609
51. Franke AA, Custer LJ, Cerna CM, Narala KK (1994) Quantitation of phytoestrogens in legumes by HPLC. J Agric Food Chem 42:1905–1913
52. Wang C, Wang L, Zhao Q, Chen J, Zheng L, Zheng M, Zhang R, Wang Z (2013) Cloud point extraction coupled with HPLC-DAD for the determination of genistein and daidzein in river water. Anal Methods 5:3688–3692
53. Oosthuizen MC, Steyn B, Theron J, Cosette P, Lindsay D, von Holy A, Brözel VS. Proteomic analysis reveals differential protein expression by *Bacillus cereus* during biofilm formation. Appl Environ Microbiol. 2002;68:2770–80.
54. Biswas S, Biswas I. Role of HtrA in surface protein expression and biofilm formation by *Streptococcus mutans*. Infect Immunity. 2005;73:6923–34.
55. Laupland, K. B. et al. Severe bloodstream infections: a population-based assessment. *Crit. Care Med. 32*, 992–997 (2004).
56. Atencia, J. & Beebe, D. J. Controlled microfluidic interfaces. *Nature 437*, 648–655 (2005).
57. Domenech, M. et al. Cellular observations enabled by microculture: paracrine signaling and population demographics. *Integr. Biol. (Camb) 1*, 267–274 (2009).
58. Young, E. W. & Beebe, D. J. Fundamentals of microfluidic cell culture in controlled microenvironments. *Chem. Soc. Rev. 39*, 1036–1048 (2010).
59. Boedicker, J. Q., Vincent, M. E. & Ismagilov, R. F. Microfluidic confinement of single cells of bacteria in small volumes initiates high-density behavior of quorum sensing and growth and reveals its variability. *Angew. Chem. Int. Ed. Engl. 48*, 5908–5911 (2009).
60. Andersen, M. R. et al. Accurate prediction of secondary metabolite gene clusters in filamentous fungi. *Proc. Natl Acad. Sci. USA 110*, E99–E107 (2013).
61. Shimizu, K. & Keller, N. P. Genetic involvement of a cAMP-dependent protein kinase in a G protein signaling pathway regulating morphological and chemical transitions in Aspergillus nidulans. *Genetics 157*, 591–600 (2001).
62. Röst, H. L. et al. OpenMS: a flexible open-source software platform for mass spectrometry data analysis. *Nat. Methods 13*, 741–748 (2016).
63. Huan, T. et al. Systems biology guided by XCMS Online metabolomics. *Nat. Methods 14*, 461–462 (2017).
64. Smith, C. A., Want, E. J., O'Maille, G., Abagyan, R. & Siuzdak, G. XCMS: processing mass spectrometry data for metabolite profiling using nonlinear peak alignment, matching, and identification. *Anal. Chem. 78*, 779–787 (2006).
65. Madji Hounoum B, Blasco H, Emond P, Mavel S. Liquid chromatography-high-resolution mass spectrometry-based cell metabolomics: experimental design, recommendations, and applications. TrAC Trends Anal Chem. 2016;75:118–28.
66. Takenaka M, Yoshida T, Hori Y, Bamba T, Mochizuki M, Vavricka CJ, et al. An ion-pair free LC-MS/MS method for quantitative metabolite profiling of microbial bioproduction systems. Talanta. 2021;222(September 2020):121625.
67. Guo, J. & Huan, T. Comparison of full-scan, data-dependent, and data-independent acquisition modes in liquid chromatography–mass spectrometry based untargeted metabolomics. *Anal. Chem. 92*, 8072–8080 (2020).

68. Ludwig, M. et al. Database-independent molecular formula annotation using Gibbs sampling through ZODIAC. *Nat. Mach. Intell. 2*, 629–641 (2020).
69. Dührkop, K. et al. SIRIUS 4: a rapid tool for turning tandem mass spectra into metabolite structure information. *Nat. Methods 16*, 299–302 (2019).
70. Tsugawa, H. et al. A cheminformatics approach to characterize metabolomes in stable-isotope-labeled organisms. *Nat. Methods 16*, 295–298 (2019).
71. Ernst, M. et al. MolNetEnhancer: enhanced molecular networks by integrating metabolome mining and annotation tools. *Metabolites 9*, 144 (2019).
72. Peregrin-Alvarez, J. M., Sanford, C. & Parkinson, J. The conservation and evolutionary modularity of metabolism. *Genome Biol. 10*, R63 (2009).
73. Stockinger, B., Hirota, K., Duarte, J. & Veldhoen, M. External influences on the immune system via activation of the aryl hydrocarbon receptor. *Semin. Immunol. 23*, 99–105 (2011).
74. Zelante, T. et al. Tryptophan catabolites from microbiota engage aryl hydrocarbon receptor and balance mucosal reactivity via interleukin-22. *Immunity 39*, 372–385 (2013).
75. Jaichander, P., Selvarajan, K., Garelnabi, M. & Parthasarathy, S. Induction of paraoxonase 1 and apolipoprotein A-I gene expression by aspirin. *J. Lipid Res. 49*, 2142–2148 (2008).
76. Qin, J. et al. A human gut microbial gene catalogue established by metagenomic sequencing. *Nature 464*, 59–65 (2010).
77. Amara, A. et al. Networks and graphs discovery in metabolomics data analysis and interpretation. *Front. Mol. Biosci. 9*, 1–15 (2022).
78. Frainay, C. & Jourdan, F. Computational methods to identify metabolic sub-networks based on metabolomic profiles. *Brief. Bioinform. 18*, 43–56 (2017).
79. Wang, M. et al. Sharing and community curation of mass spectrometry data with Global Natural Products Social Molecular Networking. *Nat. Biotechnol. 34*, 828–837 (2016).
80. Nothias, L.-F. et al. Feature-based molecular networking in the GNPS analysis environment. *Nat. Methods 17*, 905–908 (2020).
81. Yan, D., Cao, L., Zhou, M. & Mohimani, H. TransDiscovery: discovering biotransformation from human microbiota by integrating metagenomic and metabolomic data. *Metabolites 12*, 119 (2022).
82. Chen, L. et al. Metabolite discovery through global annotation of untargeted metabolomics data. *Nat. Methods 18*, 1377–1385 (2021).
83. Basu, S. et al. Sparse network modeling and Metscape-based visualization methods for the analysis of large-scale metabolomics data. *Bioinformatics 33*, btx012 (2017).
84. Gaquerel, E., Kuhl, C. & Neumann, S. Computational annotation of plant metabolomics profiles via a novel network-assisted approach. *Metabolomics 9*, 904–918 (2013).
85. Kuhl, C., Tautenhahn, R., Böttcher, C., Larson, T. R. & Neumann, S. CAMERA: an integrated strategy for compound spectra extraction and annotation of liquid chromatography/mass spectrometry data sets. *Anal. Chem. 84*, 283–289 (2012).
86. Senan, O. et al. CliqueMS: a computational tool for annotating in-source metabolite ions from LC-MS untargeted metabolomics data based on a coelution similarity network. *Bioinformatics 35*, 4089–4097 (2019).
87. Burgess, K. E. V., Borutzki, Y., Rankin, N., Daly, R. & Jourdan, F. MetaNetter 2: a cytoscape plugin for ab initio network analysis and metabolite feature classification. *J. Chromatogr. B 1071*, 68–74 (2017).
88. Shen, X. et al. Metabolic reaction network-based recursive metabolite annotation for untargeted metabolomics. *Nat. Commun. 10*, 1516 (2019).
89. Fernie, A. R., Trethewey, R. N., Krotzky, A. J. & Willmitzer, L. Metabolite profiling: from diagnostics to systems biology. *Nat. Rev. Mol. Cell Biol. 5*, 763–769 (2004).
90. Lu, W. et al. Metabolite measurement: pitfalls to avoid and practices to follow. *Annu. Rev. Biochem. 86*, 277–304 (2017). *A useful and comprehensive review highlighting the pitfalls encountered in metabolomics and providing guidelines for accurate metabolite measurements.*
91. Fiehn, O. et al. The metabolomics standards initiative (MSI). *Metabolomics 3*, 175–178 (2007).

92. Sumner, L. W. et al. Proposed minimum reporting standards for chemical analysis: Chemical Analysis Working Group (CAWG) Metabolomics Standards Initiative (MSI). *Metabolomics 3*, 211–221 (2007).
93. Sansone, S. A. et al. The Metabolomics Standards Initiative. *Nat. Biotechnol. 25*, 846–848 (2007). *This article highlights two standards and guidelines papers for MS and sample preparation by the Human Proteome Organization Proteomics Standardization Initiative (HUPO-PSI) and the Functional Genomics Experiment (FuGE) and describes how metabolomics standards should align to these.*
94. Spicer, R. A., Salek, R. & Steinbeck, C. A decade after the Metabolomics Standards Initiative: it's time for a revision. *Sci. Data 4*, 3 (2017).
95. Tohge, T. et al. From models to crop species: caveats and solutions for translational metabolomics. *Front. Plant Sci. 2*, 61 (2011).
96. Fernie, A. R. et al. Recommendations for reporting metabolite data. *Plant Cell 23*, 2477–2482 (2011).
97. Trutschel, D., Schmidt, S., Grosse, I. & Neumann, S. Experiment design beyond gut feeling: statistical tests and power to detect differential metabolites in mass spectrometry data. *Metabolomics 11*, 851–860 (2015).
98. Ap Rees, T. & Hill, S. A. Metabolic control analysis of plant metabolism. *Plant Cell Environ. 17*, 587–599 (1994).
99. Annesley, T. M. Ion suppression in mass spectrometry. *Clin. Chem. 49*, 1041–1044 (2003).
100. Antignac, J. P., Marchand, P., Le Bizec, B. & Andre, F. Identification of ractopamine residues in tissue and urine samples at ultra-trace level using liquid chromatography–positive electrospray tandem mass spectrometry. *J. Chromatogr. B Anal. Technol. Biomed. Life Sci. 774*, 59–66 (2002).
101. Kebarle, P. & Tang, L. From ions in solution to ions in the gas phase—the mechanism of electrospray mass spectrometry. *Anal. Chem. 65*, 972A–986A (1993).
102. Roessner-Tunali, U. et al. De novo amino acid biosynthesis in potato tubers is regulated by sucrose levels. *Plant Physiol. 133*, 683–692 (2003).
103. Buhrman, D. L., Price, P. I. & Rudewiczcor, P. J. Quantitation of SR 27417 in human plasma using electrospray liquid chromatography–tandem mass spectrometry: a study of ion suppression. *J. Am. Soc. Mass Spectrom. 7*, 1099–1105 (1996).
104. Price, E. J. et al. Metabolite database for root, tuber, and banana crops to facilitate modern breeding in understudied crops. *Plant J. 101*, 1258–1268 (2020). *Recent exemplary documentation of a metabolomics experiment that evaluated metabolite levels in crop species, providing not only an extensive database but moreover an excellent example of how to correctly investigate understudied species.*
105. Mallick, Himel; Franzosa, Eric A.; Mclver, Lauren J.; Banerjee, Soumya; Sirota-Madi, Alexandra; Kostic, Aleksandar D.; Clish, Clary B.; Vlamakis, Hera; Xavier, Ramnik J.; Huttenhower, Curtis Predictive metabolomic profiling of microbial communities using amplicon or metagenomic sequences. Nature Communications (2019). https://doi.org/10.1038/s41467-019-10927-1
106. Braga, Lucas Palma Perez; Pereira, Roberta Verciano; Martins, Layla Farage; Moura, Livia Maria Silva; Sanchez, Fabio Beltrame; Patané, José Salvatore Leister; da Silva, Aline Maria; Setubal, João Carlos. Genome-resolved metagenome and metatranscriptome analyses of thermophilic composting reveal key bacterial players and their metabolic interactions. BMC Genomics (2021). https://doi.org/10.1186/s12864-021-07957-9
107. Qiu, Jiazhang; Luo, Zhao-Qing Legionella and Coxiella effectors: strength in diversity and activity. Nature Reviews Microbiology (2017). https://doi.org/10.1038/nrmicro.2017.67
108. Gaya, Pilar; Arqués, Juan Luis; Medina, Margarita; Álvarez, Inmaculada; Landete, José Maria A New HPLC-PAD/HPLC-ESI-MS Method for the Analysis of Phytoestrogens Produced by Bacterial Metabolism. Food Analytical Methods (2015). https://doi.org/10.1007/s12161-015-0226-3

109. Yeor-Davidi, Ella; Zverzhinetsky, Marina; Krivitsky, Vadim; Patolsky, Fernando Real-time monitoring of bacterial biofilms metabolic activity by a redox-reactive nanosensors array. Journal of Nanobiotechnology (2020). https://doi.org/10.1186/s12951-020-00637-y
110. Rydzak, Thomas; Groves, Ryan A.; Zhang, Ruichuan; Aburashed, Raied; Pushpker, Rajnigandha; Mapar, Maryam; Lewis, Ian A. Metabolic preference assay for rapid diagnosis of bloodstream infections. Nature Communications (2022). https://doi.org/10.1038/s41467-022-30048-6
111. Barkal, Layla J.; Theberge, Ashleigh B.; Guo, Chun-Jun; Spraker, Joe; Rappert, Lucas; Berthier, Jean; Brakke, Kenneth A.; Wang, Clay C. C.; Beebe, David J.; Keller, Nancy P.; Berthier, Erwin Microbial metabolomics in open microscale platforms. Nature Communications (2016). https://doi.org/10.1038/ncomms10610
112. Giné, Roger; Capellades, Jordi; Badia, Josep M.; Vughs, Dennis; Schwaiger-Haber, Michaela; Alexandrov, Theodore; Vinaixa, Maria; Brunner, Andrea M.; Patti, Gary J.; Yanes, Oscar HERMES: a molecular-formula-oriented method to target the metabolome. Nature Methods (2021). https://doi.org/10.1038/s41592-021-01307-z
113. Zanella, Delphine; Liden, Tiffany; York, Jamie; Franchina, Flavio A.; Focant, Jean-François; Schug, Kevin A. Exploiting targeted and untargeted approaches for the analysis of bacterial metabolites under altered growth conditions. Analytical and Bioanalytical Chemistry (2021). https://doi.org/10.1007/s00216-021-03505-2
114. Chen, Li; Lu, Wenyun; Wang, Lin; Xing, Xi; Chen, Ziyang; Teng, Xin; Zeng, Xianfeng; Muscarella, Antonio D.; Shen, Yihui; Cowan, Alexis; McReynolds, Melanie R.; Kennedy, Brandon J.; Lato, Ashley M.; Campagna, Shawn R.; Singh, Mona; Rabinowitz, Joshua D. Metabolite discovery through global annotation of untargeted metabolomics data. Nature Methods (2021). https://doi.org/10.1038/s41592-021-01303-3
115. Sridharan, Gautham V.; Choi, Kyungoh; Klemashevich, Cory; Wu, Charmian; Prabakaran, Darshan; Pan, Long Bin; Steinmeyer, Shelby; Mueller, Carrie; Yousofshahi, Mona; Alaniz, Robert C.; Lee, Kyongbum; Jayaraman, Arul Prediction and quantification of bioactive microbiota metabolites in the mouse gut. Nature Communications (2014). https://doi.org/10.1038/ncomms6492
116. Zhou, Zhiwei; Luo, Mingdu; Zhang, Haosong; Yin, Yandong; Cai, Yuping; Zhu, Zheng-Jiang Metabolite annotation from knowns to unknowns through knowledge-guided multilayer metabolic networking. Nature Communications (2022). https://doi.org/10.1038/s41467-022-34537-6
117. Alseekh, Saleh; Aharoni, Asaph; Brotman, Yariv; Contrepois, Kévin; D'Auria, John; Ewald, Jan; C. Ewald, Jennifer; Fraser, Paul D.; Giavalisco, Patrick; Hall, Robert D.; Heinemann, Matthias; Link, Hannes; Luo, Jie; Neumann, Steffen; Nielsen, Jens; Perez de Souza, Leonardo; Saito, Kazuki; Sauer, Uwe; Schroeder, Frank C.; Schuster, Stefan; Siuzdak, Gary; Skirycz, Aleksandra; Sumner, Lloyd W.; Snyder, Michael P.; Tang, Huiru; Tohge, Takayuki; Wang, Yulan; Wen, Weiwei; Wu, Si; Xu, Guowang; Zamboni, Nicola; Fernie, Alisdair R. Mass spectrometry-based metabolomics: a guide for annotation, quantification and best reporting practices. Nature Methods (2021). https://doi.org/10.1038/s41592-021-01197-1

Chapter 4
Metabolic Regulation and Signaling in Bacteria

Vijay Soni and Aditya Upadhyay

4.1 Introduction by the Editor

Metabolic regulation and signaling are central to bacterial physiology, enabling microorganisms to adapt to fluctuating environments, coordinate cellular processes, and interact with hosts and other microbes. Unlike the relatively linear metabolic pathways often depicted in textbooks, bacterial metabolism is governed by complex networks of feedback loops, posttranslational modifications, and signal transduction systems. This chapter explores the versatile mechanisms by which bacteria regulate their metabolic activities and respond to internal and external signals, drawing on recent discoveries that span molecular biology, structural biochemistry, and microbial ecology.

One of the most compelling themes in bacterial signaling is the diversity of second messenger molecules. The study of cGAS-like enzymes in *Escherichia coli* reveals that bacteria synthesize a broader array of cyclic dinucleotides (CDNs) than previously recognized. The identification of cyclic UMP–AMP (cUMP–AMP), a hybrid purine–pyrimidine CDN, expands our understanding of nucleotide-based signaling and suggests that bacteria possess specialized pathways insulated from canonical purine-based CDNs. These findings not only deepen our knowledge of bacterial signal transduction but also open new avenues for therapeutic targeting of immune-modulatory pathways.

Posttranslational modifications (PTMs) are another layer of metabolic regulation that has gained attention in prokaryotic systems. The discovery of lysine lactylation (Kla) in *E. coli*, regulated by the enzymes YiaC and CobB, highlights the role of

V. Soni (✉)
Department of Medicine, Weill Cornell Medicine, New York, NY, USA
e-mail: vis2032@med.cornell.edu

A. Upadhyay
Hospital for Special Surgery, New York, NY, USA

V. Soni, A. Upadhyay (eds.), *Metabolism of Bacterial Infection*,
https://doi.org/10.1007/978-3-031-46895-7_4

PTMs in bacterial metabolism. Unlike its transcriptional role in eukaryotes, Kla in bacteria appears to modulate metabolic enzymes directly, influencing glycolysis and growth. This suggests that PTMs in bacteria are not merely vestigial remnants of eukaryotic complexity but are active regulators of core metabolic functions.

Evolutionary insights into bacterial metabolism are provided by the reconstruction of the metabolic network of the last bacterial common ancestor (LBCA). By analyzing anaerobic genomes and identifying conserved protein families, researchers have proposed that LBCA was an autotroph with a streamlined metabolic architecture. This ancestral network, centered around hubs like CO_2 and malonyl-ACP, offers a glimpse into the metabolic logic that shaped early bacterial evolution and provides a framework for understanding the origins of metabolic diversity.

Biofilm formation, a hallmark of bacterial community behavior, is also controlled by metabolic regulation. The discovery of intra-colony channels in *E. coli* biofilms, visualized using the Mesolens, challenges the traditional view of nutrient diffusion and suggests a structured nutrient uptake system. These channels may represent a form of metabolic infrastructure, allowing biofilms to optimize resource distribution and resist environmental stressors.

At the interface of metabolism and gene regulation, the systematic identification of metabolite–transcription factor (TF) interactions in *E. coli* reveals how intracellular metabolites can influence transcriptional networks. By integrating transcriptomic and metabolomic data, researchers have mapped interactions such as lysine–ArgR and tryptophan–SoxR, demonstrating that metabolites serve as both substrates and signals. This bidirectional communication between metabolism and gene expression underscores the dynamic nature of bacterial regulatory systems.

Signal transduction through two-component systems (TCSs) is another critical mechanism by which bacteria sense and respond to environmental changes. The structural elucidation of the extracellular receptor domain of WalK in *Staphylococcus aureus* reveals how small molecules can modulate TCS activity. The identification of 2,4-dihydroxybenzophenone (DHBP) as an activator of WalKR suggests that targeting extracellular domains may offer new strategies for controlling bacterial virulence.

Finally, the regulation of biofilm formation in *Vibrio cholerae* through sugar-mediated modulation of a c-di-GMP phosphodiesterase (PdeS) illustrates how metabolic inputs influence lifestyle transitions. The phosphorylation state of EIIA Glc, a component of the phosphotransferase system (PTS), determines its interaction with PdeS, thereby controlling c-di-GMP levels and biofilm dynamics.

Together, these studies reveal that bacterial metabolic regulation is a highly integrated and adaptive system, capable of responding to diverse stimuli and coordinating complex behaviors. This chapter provides a comprehensive overview of the molecular mechanisms that underpin bacterial signaling and metabolic control, setting the stage for deeper exploration of their roles in pathogenesis, evolution, and biotechnology.

4.2 Machine Generated Summaries

Disclaimer: The summaries in this chapter were generated from Springer Nature publications using abstractive AI auto-summarization: An abstraction-based summarizer creates new text based on deep learning. New phrases are created to summarize the content. As the constituted sentences are machine selected, they may not fully reflect the body of the work, so we strongly advise that the original content is read and cited. The auto-generated summaries were curated by the editor to meet Springer Nature publication standards.

Bacterial cGAS-Like Enzymes Synthesize Diverse Nucleotide Signals [76]

This is a machine-generated summary of:

Whiteley, Aaron T.; Eaglesham, James B.; de Oliveira Mann, Carina C.; Morehouse, Benjamin R.; Lowey, Brianna; Nieminen, Eric A.; Danilchanka, Olga; King, David S.; Lee, Amy S. Y.; Mekalanos, John J.; Kranzusch, Philip J. Bacterial cGAS-Like Enzymes Synthesize Diverse Nucleotide Signals [76].

Published in: Nature (2019).

Link to original: https://doi.org/10.1038/s41586-019-0953-5

If you want to cite the papers, please refer to the original.

For technical reasons we could not place the page where the original quote is coming from.

Research Highlights

Cells use second messenger molecules to amplify signals and control responses, a process crucial for immune functions. In human cells, cyclic GMP–AMP synthase (cGAS) is activated by mislocalized double-stranded DNA to produce cyclic GMP–AMP (cGAMP), which triggers immune responses. While synthetic cyclic dinucleotide (CDN) analogues are being explored for therapeutic purposes, our understanding of bacterial CDN signaling, particularly beyond purine-based CDNs, remains limited. Now, a new study by Aaron T. Whiteley and colleagues uncovers a novel pyrimidine-containing CDN synthesized by the enzyme CdnE in *Escherichia coli*, expanding the known diversity of bacterial signaling molecules.

The study employed a systematic approach to explore the diversity of products synthesized by microbial synthases related to cGAS. The researchers identified an operon in *E. coli* containing a gene encoding CdnE, hypothesized to be an alternative CDN synthase. By incubating purified CdnE with radiolabeled nucleotides and analyzing the reaction products through thin-layer chromatography, mass spectrometry, and NMR, they discovered that CdnE synthesizes cyclic UMP–AMP (cUMP–AMP), a hybrid purine–pyrimidine CDN. Structural analysis revealed that CdnE shares a nucleotidyltransferase fold with cGAS and DncV but has unique active-site features allowing it to produce cUMP–AMP. The study also demonstrated that cUMP–AMP specifically activates an adjacent phospholipase gene product in *E. coli*, suggesting a distinct signaling pathway insulated from other CDNs.

This research advances our understanding of bacterial nucleotide signaling by identifying a new class of CDNs that includes pyrimidine bases. The discovery of cUMP–AMP highlights the potential for diverse bacterial CDNs to modulate host immune responses through different receptors. Future studies could explore the ecological roles of these CDNs in bacteria–host interactions and their potential applications in biotechnology and medicine. The findings suggest that bacterial enzymes like CdnE could be harnessed to develop novel therapeutic agents targeting specific immune pathways without triggering broad immune activation.

YiaC and CobB Regulate Lysine Lactylation in Escherichia coli *[77]*

This is a machine-generated summary of:

Dong, Hanyang; Zhang, Jianji; Zhang, Hui; Han, Yue; Lu, Congcong; Chen, Chen; Tan, Xiaoxia; Wang, Siyu; Bai, Xue; Zhai, Guijin; Tian, Shanshan; Zhang, Tao; Cheng, Zhongyi; Li, Enmin; Xu, Liyan; Zhang, Kai YiaC and CobB Regulate Lysine Lactylation in *Escherichia coli* [77].

Published in: Nature Communications (2022).

Link to original: https://doi.org/10.1038/s41467-022-34399-y

If you want to cite the papers, please refer to the original.

For technical reasons we could not place the page where the original quote is coming from.

Research Highlights

Posttranslational modifications (PTMs) are crucial for regulating protein functions and influencing various cellular processes and disease progressions in eukaryotes. Recent evidence suggests that PTMs, such as lysine acetylation, also play significant roles in prokaryotes by regulating bacterial virulence, chemotaxis, and protein stability. Despite these insights, the understanding of PTMs in prokaryotes remains limited. Now, a new study by Hanyang Dong and colleagues explores lysine lactylation (Kla), a recently identified PTM in eukaryotes, to determine its existence and function in prokaryotes.

The study profiled the Kla proteome in *Escherichia coli* and identified YiaC as a lactylase and CobB as a delactylase through screening strains with overexpressed N-acetyltransferase family proteins and CobB. Using stable isotope labeling by amino acids in cell culture-based quantitative proteomics, the researchers identified endogenous Kla sites potentially regulated by YiaC and CobB. They demonstrated that YiaC catalyzes the addition of Kla while CobB removes it, both in vitro and intracellularly. The study further revealed that lactylated proteins are primarily involved in metabolism, with CobB erasing PykF to enhance glycolysis and promote bacterial growth.

This research advances our understanding of PTMs in prokaryotes by identifying a regulatory enzyme system for Kla in bacteria. The findings suggest that Kla has an extensive regulatory effect on bacterial metabolism, distinct from its role in transcription regulation in eukaryotes. Future study could explore other potential lactylases and delactylases within prokaryotic systems to further elucidate the diverse functions of Kla. Additionally, understanding how metabolites influence Kla levels could provide insights into the broader implications of PTMs on cellular processes and bacterial growth dynamics.

The Metabolic Network of the Last Bacterial Common Ancestor [78]

This is a machine-generated summary of:

Xavier, Joana C.; Gerhards, Rebecca E.; Wimmer, Jessica L. E.; Brueckner, Julia; Tria, Fernando D. K.; Martin, William F. The Metabolic Network of the Last Bacterial Common Ancestor [78].

Published in: Communications Biology (2021).

Link to original: https://doi.org/10.1038/s42003-021-01918-4

If you want to cite the papers, please refer to the original.

For technical reasons we could not place the page where the original quote is coming from.

Research Highlights

Bacteria, the most abundant and diverse domain of life, are believed to have originated from a last universal common ancestor (LUCA) that predated the divergence of bacteria and archaea. The study of early bacterial evolution is complicated by lateral gene transfer (LGT), which obscures the evolutionary history of metabolic pathways. Now, a new study by Joana C. Xavier and colleagues reconstructs the core metabolic network of the last bacterial common ancestor (LBCA) by analyzing anaerobic genomes and identifying genes widely distributed among them.

The researchers began by selecting reference genomes from anaerobic bacteria, excluding those with oxygen reductases to focus on ancient lineages. They identified protein families present across major taxonomic groups, which were then manually annotated for functional insights into LBCA's physiology. The study revealed that LBCA was likely an autotroph with a nearly complete gluconeogenetic pathway, capable of generating essential metabolites using a minimal set of reactions. The analysis also highlighted Clostridia as the modern bacterial group most similar to LBCA, suggesting it was one of the earliest diverging lineages.

The findings provide a conservative yet comprehensive view of LBCA's metabolic network, indicating it was structured around key metabolic hubs like CO_2/ACP/malonyl-ACP. Future studies could explore the role of multifunctional enzymes in early prokaryotic evolution and investigate how these findings align with geological evidence of early Earth environments. The research underscores the importance of integrating phylogenomic data with metabolic network analysis to unravel the complexities of ancient microbial life.

Intra-Colony Channels in E. coli *Function as a Nutrient Uptake System [79]*

This is a machine-generated summary of:

Rooney, Liam M.; Amos, William B.; Hoskisson, Paul A.; McConnell, Gail Intra-Colony Channels in *E. coli* Function as a Nutrient Uptake System [79].

Published in: The ISME Journal: Multidisciplinary Journal of Microbial Ecology (2020).
Link to original: https://doi.org/10.1038/s41396-020-0700-9

If you want to cite the papers, please refer to the original.
For technical reasons we could not place the page where the original quote is coming from.

Research Highlights
Bacteria often form biofilms, which are surface-attached communities enveloped in a protective extracellular matrix. These biofilms can be composed of one or more species and are found in diverse ecological niches, providing resistance to environmental stressors and antibiotics. Despite their prevalence, the ecological benefits of biofilm formation and their role in antimicrobial resistance are not fully understood. Now, a new study by Liam M. Rooney and colleagues explores the internal architecture of *Escherichia coli* macro-colony biofilms using the Mesolens, revealing a previously undocumented channel system that facilitates nutrient uptake.

Rooney and colleagues employed the Mesolens, a high-resolution imaging tool, to investigate the structure of mature *E. coli* biofilms. The study involved designing a custom imaging chamber to grow bacterial colonies under controlled conditions. The researchers used widefield epi-fluorescence mesoscopy and confocal laser-scanning mesoscopy to visualize the biofilms' internal structures. They discovered channel-like formations within the biofilms that appeared as non-fluorescing regions lined by bacterial cells. These channels were shown to facilitate the transport of nutrients and other substances into the biofilm, challenging the traditional view that nutrient uptake occurs solely through passive diffusion.

The identification of intra-colony channels represents a significant advancement in understanding biofilm architecture and function. These channels may serve as a novel nutrient acquisition system, potentially influencing how microbial communities acquire resources in various environments. Future studies could explore whether these structures are present in other bacterial species or environmental conditions, offering insights into microbial ecology and potential applications for delivering antimicrobial agents through these channels. The use of the Mesolens has proven invaluable for this research, suggesting its broader applicability in studying microbial communities and their ecological roles.

Systematic Identification of Metabolites Controlling Gene Expression in E. coli *[80]*

This is a machine-generated summary of:

Lempp, Martin; Farke, Niklas; Kuntz, Michelle; Freibert, Sven Andreas; Lill, Roland; Link, Hannes Systematic Identification of Metabolites Controlling Gene Expression in *E. coli* [80].

Published in: Nature Communications (2019).

Link to original: https://doi.org/10.1038/s41467-019-12474-1

If you want to cite the papers, please refer to the original.

For technical reasons we could not place the page where the original quote is coming from.

Research Highlights

Transcriptional regulation of metabolism is a well-characterized process, focusing on how transcription modulates enzyme abundance and, consequently, metabolic flux and metabolites. Metabolites can also convey information back to the transcription network by interacting with transcription factors (TFs), as seen in *Escherichia coli* where arginine regulates ArgR activity. However, the interactions between intracellular metabolites and TFs remain largely unexplored due to the lack of methods to identify functionally relevant metabolite–TF interactions in vivo. Now, a new study by Martin Lempp and colleagues addresses this gap by integrating transcriptome and metabolome data from *E. coli* during dynamic transitions, generating hypotheses about metabolite–TF interactions with potential functional relevance.

The study involved switching *E. coli* cultures between growth and carbon starvation phases using a bioreactor setup. Researchers measured changes in metabolite concentrations via LC-MS/MS and transcript levels through RNA-sequencing during these transitions. By correlating metabolomics data with transcriptional activity inferred from gene expression profiles, they identified potential regulatory interactions between metabolites and TFs. The team constructed a literature-based metabolite-TF network for *E. coli* and validated five predicted interactions using in vitro binding assays: lysine-ArgR, tyrosine-TrpR, glutamate-SgrR, tryptophan-SoxR, and dihydroxyacetone phosphate-DhaR. This approach demonstrated that integrating multi-omics data could reveal physiologically meaningful metabolite–TF interactions.

The study's main advance is its large-scale approach to identifying metabolite–TF interactions, which could significantly enhance our understanding of metabolic regulation dynamics. Future studies could expand this method to include two-component systems by measuring exo-metabolomes or improve TF activity estimates using fluorescent reporters. The challenge of false positives due to similar metabolite dynamics might be addressed by localized metabolic perturbations using techniques like CRISPR interference. Ultimately, mapping these interactions could lead to dynamic metabolic models and novel biotechnological applications by exploiting genetic-metabolic feedback mechanisms for strain engineering or production optimization.

Structure and Mechanism of the Essential Two-Component Signal-Transduction System WalKR in* Staphylococcus aureus *[81]

This is a machine-generated summary of:

Ji, Quanjiang; Chen, Peter J.; Qin, Guangrong; Deng, Xin; Hao, Ziyang; Wawrzak, Zdzislaw; Yeo, Won-Sik; Quang, Jenny Winjing; Cho, Hoonsik; Luo, Guan-Zheng; Weng, Xiaocheng; You, Qiancheng; Luan, Chi-Hao; Yang, Xiaojing; Bae, Taeok; Yu, Kunqian; Jiang, Hualiang; He, Chuan Structure and Mechanism of the Essential Two-Component Signal-Transduction System WalKR in *Staphylococcus aureus* [81].

Published in: Nature Communications (2016).

Link to original: https://doi.org/10.1038/ncomms11000

If you want to cite the papers, please refer to the original.

For technical reasons we could not place the page where the original quote is coming from.

Research Highlights

Signal transduction through two-component systems (TCSs) is a crucial mechanism by which bacteria adapt to environmental changes. In *Staphylococcus aureus*, the WalKR TCS is essential for regulating various cellular processes, including cell wall metabolism and virulence factor expression. Despite extensive research on the intracellular mechanisms of WalKR, the role of its extracellular receptor domain (erWalK) in signal recognition and kinase activity modulation remains poorly understood. Now, a new study by Quanjiang Ji and colleagues reveals the crystal structure of erWalK, highlighting its potential as a target for small molecules to modulate WalKR activity and treat *S. aureus* infections.

The study employed structural biology techniques to determine the crystal structure of erWalK, revealing a canonical Per-Arnt-Sim (PAS) domain involved in signal recognition. The researchers used structure-based virtual screening to identify small molecules that could bind to erWalK and modulate its function. They discovered a small molecule, 2,4-dihydroxybenzophenone (DHBP), which activates the WalKR TCS by targeting erWalK. Genetic studies involving single amino acid substitutions in erWalK demonstrated its critical role in signal transduction, as mutations led to impaired WalKR activity and reduced virulence in *S. aureus*.

The findings of this study advance our understanding of the extracellular signaling mechanisms in *S. aureus* and highlight erWalK as a promising target for therapeutic intervention. Future studies could focus on designing more potent small molecules based on DHBP's scaffold to modulate WalKR activity with higher specificity and efficacy. Additionally, exploring the interactions between erWalK and other potential ligands may provide further insights into the regulation of this essential TCS in *S. aureus*. These efforts could pave the way for developing novel anti-infective strategies against drug-resistant strains of *S. aureus* by targeting its signal transduction pathways.

Sugar-Mediated Regulation of a c-di-GMP Phosphodiesterase in **Vibrio cholerae** *[82]*

This is a machine-generated summary of:

Heo, Kyoo; Park, Young-Ha; Lee, Kyung-Ah; Kim, Joonwon; Ham, Hyeong-In; Kim, Byung-Gee; Lee, Won-Jae; Seok, Yeong-Jae Sugar-Mediated Regulation of a c-di-GMP Phosphodiesterase in *Vibrio cholerae* [82].

Published in: Nature Communications (2019).

Link to original: https://doi.org/10.1038/s41467-019-13353-5

If you want to cite the papers, please refer to the original.

For technical reasons we could not place the page where the original quote is coming from.

Research Highlights

Most bacterial species, including the pathogen *Vibrio cholerae*, can form biofilms, which are multicellular communities embedded in a self-produced polymeric matrix. This ability facilitates bacterial colonization and persistence in hosts by evading immune responses and increasing resistance to antimicrobials. Despite previous studies suggesting that the phosphoenolpyruvate phosphotransferase system (PTS) influences biofilm formation, the precise regulatory mechanisms remain unclear. Now, a new study by Kyoo Heo and colleagues explores how EIIA Glc, a component of the PTS, modulates biofilm formation in *V. cholerae* by interacting with a c-di-GMP phosphodiesterase (PdeS), depending on its phosphorylation state.

The researchers conducted ligand fishing experiments to identify proteins interacting with EIIA Glc in *V. cholerae*. They discovered that EIIA Glc interacts with PdeS, an EAL domain-containing protein involved in c-di-GMP metabolism. The study revealed that the interaction between EIIA Glc and PdeS is dependent on the phosphorylation state of EIIA Glc: phosphorylated EIIA Glc enhances PdeS activity, leading to decreased c-di-GMP levels and reduced biofilm formation, while dephosphorylated EIIA Glc inhibits PdeS activity, increasing c-di-GMP levels and promoting biofilm formation. These findings were supported by in vitro assays showing that phosphorylated EIIA Glc increases PdeS's phosphodiesterase activity, while dephosphorylated EIIA Glc inhibits it. The study also demonstrated that this regulatory mechanism affects *V. cholerae*'s ability to colonize the *Drosophila melanogaster* intestine.

This study advances our understanding of how *V. cholerae* regulates biofilm formation in response to environmental carbon sources through the PTS system. By elucidating the interaction between EIIA Glc and PdeS, it highlights a sophisticated mechanism by which *V. cholerae* modulates its lifestyle transitions between motile and sessile states. Future studies could explore how this regulatory pathway influences other phenotypes related to pathogenesis and environmental adaptation in *V. cholerae* and other bacteria with similar systems.

4.3 Conclusion

The studies presented in this chapter collectively highlight the complexity and versatility of bacterial metabolic regulation and signaling. Far from being simple organisms governed by linear biochemical pathways, bacteria exhibit a remarkable capacity to integrate environmental cues, internal metabolic states, and structural adaptations into coordinated responses that ensure survival, growth, and ecological success. This chapter has highlighted how bacteria use diverse molecular strategies—from second messengers and posttranslational modifications to transcriptional feedback and structural innovations—to regulate their metabolic networks and signal transduction systems.

One of the most striking revelations is the expanding diversity of bacterial second messengers. The discovery of cyclic UMP–AMP (cUMP–AMP) synthesized by the enzyme CdnE in *E. coli* challenges the long-held assumption that bacterial cyclic dinucleotides (CDNs) are limited to purine-based molecules. This hybrid purine–pyrimidine CDN not only broadens the chemical repertoire of bacterial signaling but also suggests the existence of specialized pathways that may interact with distinct receptors or regulatory proteins. Such findings have profound implications for understanding host–pathogen interactions, as bacterial CDNs are known to modulate immune responses. The potential to harness or inhibit these pathways for therapeutic purposes represents an exciting frontier in microbiology and immunology.

Posttranslational modifications (PTMs), once considered rare in prokaryotes, are now recognized as key regulators of bacterial metabolism. The identification of lysine lactylation (Kla) and its regulation by YiaC and CobB in *E. coli* reveals that PTMs can directly influence metabolic enzyme activity and cellular growth. Unlike their transcriptional roles in eukaryotes, bacterial PTMs appear to act as metabolic switches, fine-tuning enzymatic functions in response to intracellular metabolite levels. This insight opens new avenues for exploring how metabolic fluxes are dynamically controlled in bacterial cells and how PTMs might be exploited to modulate bacterial behavior.

The reconstruction of the metabolic network of the last bacterial common ancestor (LBCA) offers a rare glimpse into the evolutionary origins of bacterial metabolism. By identifying conserved protein families across anaerobic lineages, researchers have proposed that LBCA was an autotroph with a streamlined metabolic architecture centered around key hubs like CO_2 and malonyl-ACP. These findings not only inform our understanding of early microbial life but also provide a framework for studying metabolic evolution and the emergence of complex regulatory systems.

Biofilm formation, a critical aspect of bacterial ecology and pathogenesis, is shown to be linked to metabolic regulation. The discovery of intra-colony channels in *E. coli* biofilms reveals a structured nutrient uptake system that challenges traditional models of passive diffusion. These channels may represent a form of metabolic infrastructure, allowing biofilms to optimize resource distribution and resist environmental stress. Understanding these structures could lead to novel strategies for disrupting biofilm integrity or enhancing antimicrobial delivery.

The integration of metabolomics and transcriptomics to identify metabolite–transcription factor (TF) interactions in *E. coli* exemplifies the power of multi-omics approaches in uncovering regulatory networks. By mapping interactions such as lysine–ArgR and tryptophan–SoxR, researchers have demonstrated that metabolites serve not only as substrates but also as signals that modulate gene expression. This bidirectional communication between metabolism and transcription highlights the dynamic nature of bacterial regulatory systems and suggests new possibilities for metabolic engineering and synthetic biology.

Signal transduction through two-component systems (TCSs) remains a cornerstone of bacterial adaptation. The structural characterization of the extracellular receptor domain of WalK in *Staphylococcus aureus* and the identification of small molecule modulators like DHBP illustrate how extracellular signals can influence intracellular responses. Targeting such domains offers a promising strategy for controlling bacterial virulence and combating drug-resistant pathogens.

Finally, the regulation of biofilm formation in *V. cholerae* through sugar-mediated modulation of c-di-GMP levels demonstrates how metabolic inputs can govern lifestyle transitions. The phosphorylation-dependent interaction between EIIA Glc and PdeS exemplifies the nuanced control bacteria exert over their behavior in response to environmental carbon sources.

In conclusion, bacterial metabolic regulation and signaling are deeply interconnected systems that regulate virtually every aspect of microbial life. The studies in this chapter reveal a landscape of molecular innovation and adaptability, offering insights into bacterial physiology, evolution, and potential therapeutic targets. As research continues to uncover new layers of complexity, our understanding of bacterial metabolism will not only deepen but also inform the development of novel strategies to manipulate microbial systems for health, industry, and environmental sustainability.

Bibliography

1. Wu, J. & Chen, Z. J. Innate immune sensing and signaling of cytosolic nucleic acids. *Annu. Rev. Immunol. 32*, 461–488 (2014).
2. Corrales, L. et al. Direct activation of STING in the tumor microenvironment leads to potent and systemic tumor regression and immunity. *Cell Rep. 11*, 1018–1030 (2015).
3. Krasteva, P. V. & Sondermann, H. Versatile modes of cellular regulation via cyclic dinucleotides. *Nat. Chem. Biol. 13*, 350–359 (2017).
4. Davies, B. W., Bogard, R. W., Young, T. S. & Mekalanos, J. J. Coordinated regulation of accessory genetic elements produces cyclic di-nucleotides for *V. cholerae* virulence. *Cell 149*, 358–370 (2012).

5. Hu, D. et al. Origins of the current seventh cholera pandemic. *Proc. Natl Acad. Sci. USA 113*, E7730–E7739 (2016).
6. Dziejman, M. et al. Comparative genomic analysis of *Vibrio cholerae*: genes that correlate with cholera endemic and pandemic disease. *Proc. Natl Acad. Sci. USA 99*, 1556–1561 (2002).
7. Severin, G. B. et al. Direct activation of a phospholipase by cyclic GMP–AMP in El Tor *Vibrio cholerae*. *Proc. Natl Acad. Sci. USA 115*, E6048–E6055 (2018).
8. Kranzusch, P. J. et al. Structure-guided reprogramming of human cGAS dinucleotide linkage specificity. *Cell 158*, 1011–1021 (2014).
9. Woodward, J. J., Iavarone, A. T. & Portnoy, D. A. c-di-AMP secreted by intracellular *Listeria monocytogenes* activates a host type I interferon response. *Science 328*, 1703–1705 (2010).
10. Burroughs, A. M., Zhang, D., Schäffer, D. E., Iyer, L. M. & Aravind, L. Comparative genomic analyses reveal a vast, novel network of nucleotide-centric systems in biological conflicts, immunity and signaling. *Nucleic Acids Res. 43*, 10633–10654 (2015).
11. McFarland, A. P. et al. Sensing of bacterial cyclic dinucleotides by the oxidoreductase RECON promotes NF-κB activation and shapes a proinflammatory antibacterial state. *Immunity 46*, 433–445 (2017).
12. Jenal, U., Reinders, A. & Lori, C. Cyclic di-GMP: second messenger extraordinaire. *Nat. Rev. Microbiol. 15*, 271–284 (2017).
13. Corrigan, R. M. & Gründling, A. Cyclic di-AMP: another second messenger enters the fray. *Nat. Rev. Microbiol. 11*, 513–524 (2013).
14. Hallberg, Z. F. et al. Hybrid promiscuous (Hypr) GGDEF enzymes produce cyclic AMP–GMP (3′, 3′-cGAMP). *Proc. Natl Acad. Sci. USA 113*, 1790–1795 (2016).
15. Nelson, J. W. et al. Control of bacterial exoelectrogenesis by c-AMP–GMP. *Proc. Natl Acad. Sci. USA 112*, 5389–5394 (2015).
16. Zhang, D. et al. Metabolic regulation of gene expression by histone lactylation. *Nature 574*, 575–580 (2019).
17. Irizarry-Caro, R. A. et al. TLR signaling adapter BCAP regulates inflammatory to reparatory macrophage transition by promoting histone lactylation. *Proc. Natl Acad. Sci. USA 117*, 30628–30638 (2020).
18. Dichtl, S. et al. Lactate and IL6 define separable paths of inflammatory metabolic adaptation. *Sci. Adv. 7*, eabg3505 (2021).
19. Cui, H. et al. Lung myofibroblasts promote macrophage profibrotic activity through lactate-induced histone lactylation. *Am. J. Respir. Cell Mol. Biol. 64*, 115–125 (2021).
20. Dong, H. et al. Protein lysine de-2-hydroxyisobutyrylation by CobB in prokaryotes. *Sci. Adv. 5*, eaaw6703 (2019).
21. Starai, V. J., Celic, I., Cole, R. N., Boeke, J. D. & Escalante-Semerena, J. C. Sir2-dependent activation of acetyl-CoA synthetase by deacetylation of active lysine. *Science 298*, 2390–2392 (2002).
22. Colak, G. et al. Identification of lysine succinylation substrates and the succinylation regulatory enzyme CobB in Escherichia coli. *Mol. Cell Proteom. 12*, 3509–3520 (2013).
23. Frye, R. A. Phylogenetic classification of prokaryotic and eukaryotic Sir2-like proteins. *Biochem. Biophys. Res. Commun. 273*, 793–798 (2000).
24. Moreno-Yruela, C. et al. Class I histone deacetylases (HDAC1-3) are histone lysine delactylases. *Sci. Adv. 8*, eabi6696 (2022).
25. Zhang, X. et al. Screening, expression, purification and characterization of CoA-transferases for lactoyl-CoA generation. *J. Ind. Microbiol. Biotechnol. 46*, 899–909 (2019).
26. Kump, L. R. The rise of atmospheric oxygen. *Nature 451*, 277–278 (2008).
27. Martin, W. F. & Sousa, F. L. Early microbial evolution: the age of anaerobes. *Cold Spring Harb. Perspect. Biol. 8*, a018127 (2016).
28. Fischer, W. W., Hemp, J. & Johnson, J. E. Evolution of oxygenic photosynthesis. *Annu. Rev. Earth Planet. Sci. 44*, 647–683 (2016).
29. Xavier, J. C., Patil, K. R. & Rocha, I. Metabolic models and gene essentiality data reveal essential and conserved metabolism in prokaryotes. *PLoS Comput. Biol. 14*, e1006556 (2018).

30. Nagies, F. S. P., Brueckner, J., Tria, F. D. K. & Martin, W. F. A spectrum of verticality across genes. *PLoS Genet. 16*, e1009200 (2020).
31. Pilhofer, M. et al. Characterization and evolution of cell division and cell wall synthesis genes in the bacterial phyla Verrucomicrobia, Lentisphaerae, Chlamydiae, and Planctomycetes and phylogenetic comparison with rRNA genes. *J. Bacteriol. 190*, 3192–3202 (2008).
32. Popa, O. & Dagan, T. Trends and barriers to lateral gene transfer in prokaryotes. *Curr. Opin. Microbiol. 14*, 615–623 (2011).
33. Nelson-Sathi, S. et al. Origins of major archaeal clades correspond to gene acquisitions from bacteria. *Nature 517*, 77–80 (2015).
34. Boucher, Y. et al. Lateral gene transfer and the origins of prokaryotic groups. *Annu. Rev. Genet. 37*, 283–328 (2003).
35. Brinkmann, H., van der Giezen, M., Zhou, Y., de Raucourt, G. P. & Philippe, H. An empirical assessment of long-branch attraction artefacts in deep eukaryotic phylogenomics. *Syst. Biol. 54*, 743–757 (2005).
36. Taib, N. et al. Genome-wide analysis of the Firmicutes illuminates the diderm/monoderm transition. *Nat. Ecol. Evol. 4*, 1661–1672 (2020).
37. Coleman, G. et al. A rooted phylogeny resolves early bacterial evolution. *bioRxiv* https://doi.org/10.1101/2020.07.15.205187 (2020).
38. Weiss, M. C. et al. The physiology and habitat of the last universal common ancestor. *Nat. Microbiol. 1*, 16116 (2016).
39. Schink, B., Thiemann, V., Laue, H. & Friedrich, M. W. *Desulfotignum phosphitoxidans* sp. nov., a new marine sulfate reducer that oxidizes phosphite to phosphate. *Arch. Microbiol. 177*, 381–391 (2002).
40. Ikeda-Ohtsubo, W. et al. '*Candidatus* Adiutrix intracellularis', an endosymbiont of termite gut flagellates, is the first representative of a deep-branching clade of *Deltaproteobacteria* and a putative homoacetogen. *Environ. Microbiol. 18*, 2548–2564 (2016).
41. Nisbet, E. & Sleep, N. H. The habitat and nature of early life. *Nature 409*, 1083–1091 (2001).
42. Rudge TJ, Federici F, Steiner PJ, Kan A, Haseloff J. Cell polarity-driven instability generates self-organized, fractal patterning of cell layers. ACS Synth Biol. 2013;2:705–14.
43. Blanchard AE, Lu T. Bacterial social interactions drive the emergence of differential spatial colony structures. BMC Syst Biol. 2015;9:59. https://doi.org/10.1186/s12918-015-0188-5.
44. Smith WPJ, Davit Y, Osborne JM, Kim W, Foster KR, Pitt-Francis JM. Cell morphology drives spatial patterning in microbial communities. Proc Natl Acad Sci. 2017;114:E280–6.
45. Goldschmidt F, Regoes RR, Johnson DR. Successive range expansion promotes diversity and accelerates evolution in spatially structured microbial populations. ISME J. 2017;11:2112.
46. Jauffred L, Vejborg RM, Korolev KS, Brown S, Oddershede LB. Chirality in microbial biofilms is mediated by close interactions between the cell surface and the substratum. ISME J. 2017;11:1688.
47. Elbing KL, Brent R. Recipes and tools for culture of *Escherichia coli*. Curr Protoc Mol Biol. 2019;125:e83.
48. Stoodley P, Debeer D, Lewandowski Z. Liquid flow in biofilm systems. Appl Environ Microbiol. 1994;60:2711–6.
49. Drury WJ, Characklis WG, Stewart PS. Interactions of 1 μm latex particles with *Pseudomonas aeruginosa* biofilms. Water Res. 1993;27:1119–26.
50. Folly, B. B. et al. Assessment of the interaction between the flux-signaling metabolite fructose-1,6-bisphosphate and the bacterial transcription factors CggR and Cra. *Mol. Microbiol. 109*, 278–290 (2018).
51. Piazza, I. et al. A map of protein-metabolite interactions reveals principles of chemical communication. *Cell 172*, 358–372.e23 (2018).
52. Kochanowski, K. et al. Few regulatory metabolites coordinate expression of central metabolic genes in Escherichia coli. *Mol. Syst. Biol. 13*, 903–903 (2017).
53. Razo-Mejia, M. et al. Tuning transcriptional regulation through signaling: a predictive theory of allosteric induction. *Cell Syst. 6*, 456–469.e10 (2018).

54. Szurmant, H., White, R. A. & Hoch, J. A. Sensor complexes regulating two-component signal transduction. *Curr. Opin. Struct. Biol. 17*, 706–715 (2007).
55. Kleerebezem, M., Quadri, L. E., Kuipers, O. P. & de Vos, W. M. Quorum sensing by peptide pheromones and two-component signal-transduction systems in Gram-positive bacteria. *Mol. Microbiol. 24*, 895–904 (1997).
56. Delaune, A. et al. The WalKR system controls major staphylococcal virulence genes and is involved in triggering the host inflammatory response. *Infect. Immun. 80*, 3438–3453 (2012).
57. Dubrac, S., Bisicchia, P., Devine, K. M. & Msadek, T. A matter of life and death: cell wall homeostasis and the WalKR (YycGF) essential signal transduction pathway. *Mol. Microbiol. 70*, 1307–1322 (2008).
58. Dubrac, S., Boneca, I. G., Poupel, O. & Msadek, T. New insights into the WalK/WalR (YycG/YycF) essential signal transduction pathway reveal a major role in controlling cell wall metabolism and biofilm formation in *Staphylococcus aureus*. *J. Bacteriol. 189*, 8257–8269 (2007).
59. Martin, P. K., Li, T., Sun, D., Biek, D. P. & Schmid, M. B. Role in cell permeability of an essential two-component system in *Staphylococcus aureus*. *J. Bacteriol. 181*, 3666–3673 (1999).
60. Kadioglu, A., Echenique, J., Manco, S., Trombe, M. C. & Andrew, P. W. The MicAB two-component signaling system is involved in virulence of *Streptococcus pneumoniae*. *Infect. Immun. 71*, 6676–6679 (2003).
61. Jansen, A. et al. Role of insertion elements and *yycFG* in the development of decreased susceptibility to vancomycin in *Staphylococcus aureus*. *Int. J. Med. Microbiol. 297*, 205–215 (2007).
62. Liu, M. et al. Defects *in ex vivo* and *in vivo* growth and sensitivity to osmotic stress of group A *Streptococcus* caused by interruption of response regulator gene *vicR*. *Microbiology. 152*, 967–978 (2006).
63. Szurmant, H., Bu, L., Brooks, C. L. 3rd & Hoch, J. A. An essential sensor histidine kinase controlled by transmembrane helix interactions with its auxiliary proteins. *Proc. Natl Acad. Sci. USA 105*, 5891–5896 (2008).
64. Szurmant, H., Mohan, M. A., Imus, P. M. & Hoch, J. A. YycH and YycI interact to regulate the essential YycFG two-component system in *Bacillus subtilis*. *J. Bacteriol. 189*, 3280–3289 (2007).
65. Bae, T. & Schneewind, O. Allelic replacement in *Staphylococcus aureus* with inducible counter-selection. *Plasmid 55*, 58–63 (2006).
66. Karatan, E. & Watnick, P. Signals, regulatory networks, and materials that build and break bacterial biofilms. *Microbiol. Mol. Biol. Rev. 73*, 310–347 (2009).
67. Hayes, C. A., Dalia, T. N. & Dalia, A. B. Systematic genetic dissection of PTS in *Vibrio cholerae* uncovers a novel glucose transporter and a limited role for PTS during infection of a mammalian host. *Mol. Microbiol. 104*, 568–579 (2017).
68. Houot, L., Chang, S., Absalon, C. & Watnick, P. I. *Vibrio cholerae* phosphoenolpyruvate phosphotransferase system control of carbohydrate transport, biofilm formation, and colonization of the germfree mouse intestine. *Infect. Immun. 78*, 1482–1494 (2010).
69. Houot, L., Chang, S., Pickering, B. S., Absalon, C. & Watnick, P. I. The phosphoenolpyruvate phosphotransferase system regulates *Vibrio cholerae* biofilm formation through multiple independent pathways. *J. Bacteriol. 192*, 3055–3067 (2010).
70. Fong, J. C. & Yildiz, F. H. Interplay between cyclic AMP-cyclic AMP receptor protein and cyclic di-GMP signaling in *Vibrio cholerae* biofilm formation. *J. Bacteriol. 190*, 6646–6659 (2008).
71. Beyhan, S., Bilecen, K., Salama, S. R., Casper-Lindley, C. & Yildiz, F. H. Regulation of rugosity and biofilm formation in *Vibrio cholerae*: comparison of VpsT and VpsR regulons and epistasis analysis of *vpsT*, *vpsR*, and *hapR*. *J. Bacteriol. 189*, 388–402 (2007).
72. Sauer, K. et al. Characterization of nutrient-induced dispersion in *Pseudomonas aeruginosa* PAO1 biofilm. *J. Bacteriol. 186*, 7312–7326 (2004).
73. Gjermansen, M., Ragas, P., Sternberg, C., Molin, S. & Tolker-Nielsen, T. Characterization of starvation-induced dispersion in *Pseudomonas putida* biofilms. *Environ. Microbiol. 7*, 894–906 (2005).

74. Schleheck, D. et al. *Pseudomonas aeruginosa* PAO1 preferentially grows as aggregates in liquid batch cultures and disperses upon starvation. *PLoS ONE 4*, e5513 (2009).
75. Liang, W., Pascual-Montano, A., Silva, A. J. & Benitez, J. A. The cyclic AMP receptor protein modulates quorum sensing, motility and multiple genes that affect intestinal colonization in *Vibrio cholerae*. *Microbiology 153*, 2964–2975 (2007).
76. Whiteley, Aaron T.; Eaglesham, James B.; de Oliveira Mann, Carina C.; Morehouse, Benjamin R.; Lowey, Brianna; Nieminen, Eric A.; Danilchanka, Olga; King, David S.; Lee, Amy S. Y.; Mekalanos, John J.; Kranzusch, Philip J. Bacterial cGAS-like enzymes synthesize diverse nucleotide signals. Nature (2019). https://doi.org/10.1038/s41586-019-0953-5
77. Dong, Hanyang; Zhang, Jianji; Zhang, Hui; Han, Yue; Lu, Congcong; Chen, Chen; Tan, Xiaoxia; Wang, Siyu; Bai, Xue; Zhai, Guijin; Tian, Shanshan; Zhang, Tao; Cheng, Zhongyi; Li, Enmin; Xu, Liyan; Zhang, Kai YiaC and CobB regulate lysine lactylation in Escherichia coli. Nature Communications (2022). https://doi.org/10.1038/s41467-022-34399-y
78. Xavier, Joana C.; Gerhards, Rebecca E.; Wimmer, Jessica L. E.; Brueckner, Julia; Tria, Fernando D. K.; Martin, William F. The metabolic network of the last bacterial common ancestor. Communications Biology (2021). https://doi.org/10.1038/s42003-021-01918-4
79. Rooney, Liam M.; Amos, William B.; Hoskisson, Paul A.; McConnell, Gail Intra-colony channels in E. coli function as a nutrient uptake system. The ISME Journal: Multidisciplinary Journal of Microbial Ecology (2020). https://doi.org/10.1038/s41396-020-0700-9
80. Lempp, Martin; Farke, Niklas; Kuntz, Michelle; Freibert, Sven Andreas; Lill, Roland; Link, Hannes Systematic identification of metabolites controlling gene expression in E. coli. Nature Communications (2019). https://doi.org/10.1038/s41467-019-12474-1
81. Ji, Quanjiang; Chen, Peter J.; Qin, Guangrong; Deng, Xin; Hao, Ziyang; Wawrzak, Zdzislaw; Yeo, Won-Sik; Quang, Jenny Winjing; Cho, Hoonsik; Luo, Guan-Zheng; Weng, Xiaocheng; You, Qiancheng; Luan, Chi-Hao; Yang, Xiaojing; Bae, Taeok; Yu, Kunqian; Jiang, Hualiang; He, Chuan Structure and mechanism of the essential two-component signal-transduction system WalKR in Staphylococcus aureus. Nature Communications (2016). https://doi.org/10.1038/ncomms11000
82. Heo, Kyoo; Park, Young-Ha; Lee, Kyung-Ah; Kim, Joonwon; Ham, Hyeong-In; Kim, Byung-Gee; Lee, Won-Jae; Seok, Yeong-Jae Sugar-mediated regulation of a c-di-GMP phosphodiesterase in Vibrio cholerae. Nature Communications (2019). https://doi.org/10.1038/s41467-019-13353-5

Chapter 5
Bacterial Nutrient Acquisition and Its Function in Diseases

Vijay Soni and Aditya Upadhyay

5.1 Introduction by the Editor

Nutrient acquisition is a fundamental aspect of bacterial survival, growth, and pathogenicity. In diverse environments—from the human gut to aquatic ecosystems—bacteria must navigate fluctuating nutrient environments, often competing with other microbes and host cells for essential resources. This chapter explores the complex strategies bacteria employ to acquire nutrients and how these processes influence microbial physiology, interspecies interactions, and disease outcomes. Through a synthesis of recent studies, the chapter highlights the dynamic interplay between nutrient availability, microbial metabolism, and host–pathogen interactions.

One of the most striking insights into bacterial cooperation under nutrient stress comes from studies of interspecies interactions. In artificial consortia, such as those formed between *Desulfovibrio vulgaris* and *Clostridium acetobutylicum*, nutrient starvation triggers direct cell–cell contact and exchange of cytoplasmic materials. This physical and metabolic coupling enables survival and enhanced metabolic output, such as increased hydrogen production. These findings challenge the traditional view of bacteria as isolated entities and underscore the importance of cooperative behavior in microbial ecosystems, particularly under stress.

Nutrient regulation also plays a critical role in shaping the metabolic behavior of gut bacteria. For instance, L-glutamine, a key amino acid in host metabolism, modulates amino acid utilization by intestinal microbes. Studies in pig-derived bacterial cultures reveal that glutamine supplementation alters the uptake and metabolism of other amino acids in a species-specific and dose-dependent manner. This suggests

V. Soni (✉)
Department of Medicine, Weill Cornell Medicine, New York, NY, USA
e-mail: vis2032@med.cornell.edu

A. Upadhyay
Hospital for Special Surgery, New York, NY, USA

V. Soni, A. Upadhyay (eds.), *Metabolism of Bacterial Infection*,
https://doi.org/10.1007/978-3-031-46895-7_5

that dietary components can fine-tune microbial metabolism, with implications for gut health, nutrient absorption, and immune modulation.

Environmental conditions such as oxygen levels and nutrient richness also influence bacterial virulence traits. In *Mycobacterium avium* subsp. *hominissuis*, eutrophic and hypoxic conditions promote biofilm-like pellicle formation, a key factor in persistence and resistance. Glycopeptidolipids (GPLs) are central to this process, with mutants deficient in GPLs forming thinner, less structured pellicles. These findings highlight how environmental cues regulate virulence-associated phenotypes and suggest potential targets for disrupting biofilm formation in pathogenic mycobacteria.

At the single-cell level, nutrient and salt depletion can synergistically enhance glucose metabolism in *Escherichia coli*. Using microfluidic devices and fluorescent glucose analogs, researchers have shown that combined stress conditions increase glucose uptake and degradation without altering gene expression. Instead, these changes are regulated at the translational and posttranslational levels, revealing a previously unrecognized metabolic control that may be exploited for biotechnological or therapeutic purposes.

The relationship between diet and immunity is another critical theme. Dietary components such as vitamins A and D, phytochemicals, and fatty acids interact with nuclear receptors in immune cells, modulating T cell differentiation and mucosal immunity. These interactions influence the balance between pro-inflammatory and regulatory responses, particularly in the gut, and may help explain the rising incidence of immune-mediated diseases in industrialized societies.

Nutrient limitation also shapes microbial social behavior. In *Pseudomonas aeruginosa*, the cost of producing siderophores like pyoverdine varies depending on which nutrient is limiting. When carbon or nitrogen is scarce, siderophore production imposes a fitness cost, influencing the balance between cooperative and cheating strains. These findings provide a physiological basis for understanding microbial cooperation and its evolutionary stability.

In aquatic systems, nutrient constraints modulate the temperature sensitivity of bacterial growth efficiency. Under nutrient-rich conditions, rising temperatures reduce growth efficiency by increasing respiration more than production. In contrast, nutrient-limited conditions buffer this effect, suggesting that nutrient availability can override temperature-driven metabolic changes. This has implications for predicting microbial responses to climate change and managing carbon cycling in freshwater ecosystems.

Finally, the concept of nutritional immunity—where hosts sequester essential metals to limit pathogen growth—illustrates the competitive dynamics at the host–pathogen interface. Bacteria counter these defenses with high-affinity uptake systems and siderophores, as seen in *Acinetobacter baumannii* and other pathogens. Understanding these mechanisms opens new avenues for antimicrobial development.

Together, these studies reveal that nutrient acquisition is not merely a survival strategy but a central determinant of microbial behavior, community dynamics, and disease progression. This chapter lays the groundwork for exploring how manipulating nutrient availability can influence microbial ecosystems and host health.

5.2 Machine Generated Summaries

Disclaimer: The summaries in this chapter were generated from Springer Nature publications using abstractive AI auto-summarization: An abstraction-based summarizer creates new text based on deep learning. New phrases are created to summarize the content. As the constituted sentences are machine selected, they may not fully reflect the body of the work, so we strongly advise that the original content is read and cited. The auto-generated summaries were curated by the editor to meet Springer Nature publication standards.

Nutritional Stress Induces Exchange of Cell Material and Energetic Coupling Between Bacterial Species [117]

This is a machine-generated summary of:

Benomar, Saida; Ranava, David; Cárdenas, María Luz; Trably, Eric; Rafrafi, Yan; Ducret, Adrien; Hamelin, Jérôme; Lojou, Elisabeth; Steyer, Jean-Philippe; Giudici-Orticoni, Marie-Thérèse Nutritional Stress Induces Exchange of Cell Material and Energetic Coupling Between Bacterial Species [117]

Published in: Nature Communications (2015)

Link to original: https://doi.org/10.1038/ncomms7283

If you want to cite the papers, please refer to the original.

For technical reasons we could not place the page where the original quote is coming from.

Research Highlights

Bacteria are often studied in isolation, yet in nature, they coexist with other microorganisms, forming complex consortia that drive essential biogeochemical cycles. These microbial networks involve interactions such as metabolite exchange, allowing bacteria to occupy diverse ecological niches and exhibit enhanced metabolic capabilities. Despite their significance, the physical and metabolic interactions within these consortia remain poorly understood. Now, a new study by Saida Benomar and colleagues explores the interactions between *Desulfovibrio vulgaris* Hildenborough and *Clostridium acetobutylicum*, revealing that nutrient starvation induces interspecies cell–cell interaction and exchange of cytoplasmic material, leading to significant metabolic changes.

The study constructed an artificial consortium of *D. vulgaris*, a sulfate-reducing bacterium, and *C. acetobutylicum*, a fermentative bacterium. The researchers cultivated these bacteria in a defined medium that supported *C. acetobutylicum* growth but only allowed *D. vulgaris* survival under starvation conditions. Through various experimental approaches including flow cytometry, quantitative PCR, and fluorescence microscopy, the team demonstrated that physical contact between the two species is necessary for *D. vulgaris* growth in co-culture. The study found that this interaction facilitates the exchange of molecules such as calcein and proteins like GFP and mCherry between the bacteria. Notably, this exchange was absent when the bacteria were separated by a dialysis membrane, indicating that direct contact is crucial for material transfer.

The findings suggest that nutrient stress triggers physical interactions between bacterial species, enabling them to share resources and adapt to challenging environments. This interaction not only supports *D. vulgaris* growth but also enhances hydrogen production through altered metabolic pathways in *C. acetobutylicum*. Future studies could explore how these insights might be applied to ecological engineering or biotechnology to manipulate microbial ecosystems without genetic modification. The research highlights a novel form of bacterial cooperation that could inform strategies for improving industrial processes involving microbial consortia.

l-Glutamine Regulates Amino Acid Utilization by Intestinal Bacteria [118]

This is a machine-generated summary of:

Dai, Zhao-Lai; Li, Xi-Long; Xi, Peng-Bin; Zhang, Jing; Wu, Guoyao; Zhu, Wei-Yun l-Glutamine Regulates Amino Acid Utilization by Intestinal Bacteria [118]

Published in: Amino Acids (2012)

Link to original: https://doi.org/10.1007/s00726-012-1264-4

If you want to cite the papers, please refer to the original.

For technical reasons we could not place the page where the original quote is coming from.

Research Highlights

L-Glutamine, the most abundant free alpha-amino acid in the body, plays a crucial role in whole-body nutrient metabolism and health. It is a major fuel for small-intestinal mucosal cells, maintaining the integrity and function of the small intestine by regulating gene expression, protein turnover, immune function, cell proliferation, and apoptosis. Despite its importance, the mechanisms by which glutamine affects amino acid (AA) utilization by luminal bacteria in the small intestine remain unclear. Now, a new study by Zhao-Lai Dai and colleagues explores how glutamine regulates bacterial metabolism of AAs in pig small-intestinal bacteria, potentially reducing AA utilization.

The study employed cultures of small-intestinal bacteria derived from pigs, including *Streptococcus* sp., *Escherichia coli*, *Klebsiella sp.*, jejunal mixed bacteria, and ileal mixed bacteria. These cultures were maintained in an anaerobic semi-defined medium with varying concentrations of glutamine. The researchers measured bacterial growth and AA concentrations before and after incubation to assess the effects of glutamine on AA utilization. The results showed that glutamine's addition reduced net utilization of certain AAs like glutamate while increasing others like citrulline and ornithine in specific bacterial strains. Statistical analyses confirmed that these effects were dose-dependent and varied across different bacterial species.

The findings suggest that glutamine plays a regulatory role in the metabolism of AAs by small-intestinal bacteria, which could be crucial for maintaining gut health and function. The study highlights the potential for dietary glutamine supplementation to modulate AA utilization by both host and gut bacteria, thereby supporting gut integrity. Future studies could focus on uncovering specific mechanisms through which glutamine influences AA metabolism in intestinal bacteria and its implications for gut health. Additionally, understanding these interactions may lead to new strategies for modulating AA metabolic networks to maintain robust gut ecology and function.

Effects of Nutritional and Ambient Oxygen Condition on Biofilm Formation in **Mycobacterium avium** *subsp.* **hominissuis** *via Altered Glycolipid Expression [119]*

This is a machine-generated summary of:

Totani, Takahiro; Nishiuchi, Yukiko; Tateishi, Yoshitaka; Yoshida, Yutaka; Kitanaka, Hiromi; Niki, Mamiko; Kaneko, Yukihiro; Matsumoto, Sohkichi Effects of Nutritional and Ambient Oxygen Condition on Biofilm Formation in *Mycobacterium avium* subsp. *hominissuis* via Altered Glycolipid Expression [119]

Published in: Scientific Reports (2017)

Link to original: https://doi.org/10.1038/srep41775

If you want to cite the papers, please refer to the original.

For technical reasons we could not place the page where the original quote is coming from.

Research Highlights

Nontuberculous pathogenic mycobacteria, particularly the *Mycobacterium avium* complex (MAC), have emerged as significant etiological agents of infectious diseases, with a notable increase in cases among immunocompetent individuals in industrialized countries like the United States and Japan. This rise in MAC disease incidence has shifted focus from HIV-infected patients to those without clear cellular immunodeficiency. Despite this growing concern, there is limited understanding of the optimal conditions for biofilm formation by *Mycobacterium avium* subsp. *hominissuis* (MAH), a key factor in its persistence and virulence. Now, a new study by Takahiro Totani and colleagues explores the environmental conditions that promote biofilm-like pellicle formation in MAH, revealing that eutrophic and hypoxic conditions are crucial for this process.

The study involved analyzing the effects of various nutritional and gaseous conditions on pellicle formation by four MAH strains, including environmental isolates from patients' bathrooms. The researchers found that pellicle formation was enhanced under eutrophic and hypoxic conditions but not under oligotrophic or normoxic conditions. They also investigated the role of glycopeptidolipids (GPLs) in pellicle development by comparing wild-type strains with GPL-deficient rough mutants. The wild-type strains formed thick pellicles at the air–liquid interface, while the rough mutants produced thinner pellicles, suggesting that GPLs play a significant role in pellicle development. Additionally, scanning electron microscopy revealed ultrastructural differences between wild-type strains and rough mutants, further highlighting the importance of GPLs in maintaining cell distribution and integrity within pellicles.

The findings from Totani et al.'s study provide new insights into the ecological adaptation of MAH through biofilm-like pellicle formation under specific environmental conditions. The research underscores the critical role of GPLs in this process and suggests that other factors may also contribute to pellicle formation, as even rough mutants formed biocide-resistant pellicles. Future studies could focus on elucidating these additional factors and exploring potential strategies for controlling MAH infections by targeting biofilm formation mechanisms. These insights could inform better hygiene practices and pharmaceutical developments aimed at mitigating MAC disease risks.

Nutrient and Salt Depletion Synergistically Boosts Glucose Metabolism in Individual Escherichia coli *Cells [120]*

This is a machine-generated summary of:

Glover, Georgina; Voliotis, Margaritis; Łapińska, Urszula; Invergo, Brandon M.; Soanes, Darren; O'Neill, Paul; Moore, Karen; Nikolic, Nela; Petrov, Peter G.;

Milner, David S.; Roy, Sumita; Heesom, Kate; Richards, Thomas A.; Tsaneva-Atanasova, Krasimira; Pagliara, Stefano Nutrient and Salt Depletion Synergistically Boosts Glucose Metabolism in Individual *Escherichia coli* Cells [120]

Published in: Communications Biology (2022)
Link to original: https://doi.org/10.1038/s42003-022-03336-6

If you want to cite the papers, please refer to the original.

For technical reasons we could not place the page where the original quote is coming from.

Research Highlights

All core metabolic networks require the uptake and utilization of carbon sources, which is crucial for the growth and survival of heterotrophic organisms in nutrient-scarce environments. In bacteria, sugars like glucose are vital for powering replication, making storage compounds, and producing secondary metabolites. While glucose uptake and metabolism have been extensively studied in optimal or low-nutrient conditions, the effects of simultaneous environmental changes such as salinity variations remain less understood. Now, a new study by Georgina Glover and colleagues explores how simultaneous nutrient and salinity depletion affects glucose metabolism in *Escherichia coli*, revealing that these conditions synergistically enhance glucose uptake and degradation.

The study utilized a microfluidic mother machine device to introduce a clonal *E. coli* population to varying environmental conditions. By employing fluorescent glucose analogs like -NBDG, the researchers quantified glucose uptake and accumulation at the single-cell level under different salinity and nutrient conditions. They found that while nutrient or salinity depletion alone had distinct effects on glucose accumulation, their combination led to a significant increase in both uptake and degradation rates. This was confirmed through flow cytometry measurements and colorimetric assays. The study also demonstrated that these metabolic changes were not due to cell size variations or transcriptional changes but were instead regulated at the translational and posttranslational levels.

The findings highlight a previously unrecognized phenomenon where combined nutrient and salinity depletion enhances bacterial glucose metabolism beyond what either condition alone can achieve. This suggests that bacteria can adapt their metabolic processes through translational modifications rather than changes in gene expression. Future studies could explore how these insights might inform biotechnological applications such as optimizing bacterial production processes or developing new antimicrobial strategies. Additionally, understanding these mechanisms could have implications for food preservation techniques where salt is used as an antimicrobial agent, as well as for medical conditions involving fluctuating environmental conditions like cystic fibrosis-associated lung infections.

Dietary Influences on Intestinal Immunity [121]

This is a machine-generated summary of:

Veldhoen, Marc; Brucklacher-Waldert, Verena Dietary Influences on Intestinal Immunity [121]

Published in: Nature Reviews Immunology (2012)

Link to original: https://doi.org/10.1038/nri3299

If you want to cite the papers, please refer to the original.

For technical reasons we could not place the page where the original quote is coming from.

Research Highlights

In the nineteenth century, Jean Anthelme Brillat-Savarin famously stated, "Dis-moi ce que tu manges, je te dirai ce que tu es," highlighting the profound connection between diet and health. Modern research supports this notion, linking dietary habits to various disorders, including cardiovascular diseases, metabolic conditions, and immune disorders such as diabetes and multiple sclerosis. Despite identifying several susceptibility genes for these conditions, each gene contributes only modestly to disease heritability. The increasing prevalence of chronic inflammatory disorders in industrialized regions suggests that environmental factors like diet may play a significant role. Now, a new study by Marc Veldhoen and colleagues explores the intricate relationship between diet and immune function, focusing on how dietary components interact with nuclear receptors in immune cells.

The study delves into the mechanisms by which dietary compounds influence immune responses through nuclear receptors such as RARs, RXRs, and AHR. These receptors act as transcription factors that mediate the effects of vitamins A and D, phytochemicals, and fatty acids on immune cells. The researchers highlight how these dietary components can modulate the differentiation and function of T cells and innate lymphoid cells (ILCs) at mucosal sites like the intestine. For instance, vitamin A metabolites are shown to enhance regulatory T cell (T Reg) differentiation while inhibiting pro-inflammatory T helper (T H) cell responses. Similarly, vitamin D influences T cell activity by modulating dendritic cell maturation and cytokine production. The study also examines how phytochemicals from cruciferous vegetables activate AHR to maintain intestinal immune homeostasis by supporting IELs and ILCs.

The findings underscore the complexity of interactions between diet and immunity, suggesting that dietary components can significantly influence immune homeostasis through nuclear receptor pathways. Future studies could focus on elucidating the precise molecular mechanisms underlying these interactions to develop targeted dietary interventions for preventing or managing inflammatory disorders. Additionally, understanding how different dietary patterns affect nuclear receptor activity could inform public health strategies aimed at reducing the incidence of diet-related diseases. This research provides a foundation for exploring how personalized nutrition might optimize immune function and overall health.

Nutrient Limitation Determines the Fitness of Cheaters in Bacterial Siderophore Cooperation [122]

This is a machine-generated summary of:

Sexton, D. Joseph; Schuster, Martin Nutrient Limitation Determines the Fitness of Cheaters in Bacterial Siderophore Cooperation [122]

Published in: Nature Communications (2017)

Link to original: https://doi.org/10.1038/s41467-017-00222-2

If you want to cite the papers, please refer to the original.

For technical reasons we could not place the page where the original quote is coming from.

Research Highlights

Cooperative behaviors that provide collective benefits are widespread across all domains of life, from animals to bacteria. However, the evolution and maintenance of these behaviors are challenging to explain due to the presence of non-contributing cheaters who benefit without incurring costs. While several factors have been proposed to explain cooperation in animals, microbial cooperative behaviors have gained attention as model systems for testing these theories. Despite this, little focus has been given to the costs and benefits of these behaviors in the context of microbial growth physiology. Now, a new study by D. Joseph Sexton and colleagues investigates the effects of nutrient limitation on iron acquisition via pyoverdine molecules in *Pseudomonas aeruginosa*, revealing that the fitness cost of pyoverdine production depends on the type of nutrient limitation.

The study utilized a whole-genome metabolic model paired with controlled in vitro evolution experiments to explore how nutrient limitation affects the fitness cost of pyoverdine production. The researchers used flux-balance analysis to model bacterial growth and pyoverdine secretion under different nutrient limitations, such as carbon (C), nitrogen (N), phosphorus (P), sulfur (S), and iron (Fe). They found that pyoverdine production is costly when C or N is limiting because these nutrients are building blocks for pyoverdine, leading to competition between growth and secretion. In contrast, when P, S, or Fe is limiting, pyoverdine production incurs no fitness cost as its building blocks are in relative excess. These predictions were confirmed using chemostat cultures that allow precise control over nutrient availability, unlike batch cultures where dynamic changes in media composition can obscure results.

The findings have significant implications for understanding the evolutionary stability of cooperative behaviors like siderophore secretion. The study highlights that the intrinsic molecular composition of a secreted product determines fitness costs and cooperator-cheater dynamics based on resource availability. Future studies could explore how these principles apply to other microbial systems and environments, such as marine ecosystems or biotechnological applications. The research underscores the importance of considering ecophysiological context in social evolution studies and suggests that fitness costs may vary more widely than currently appreciated.

Nutrient Constraints on Metabolism Affect the Temperature Regulation of Aquatic Bacterial Growth Efficiency [123]

This is a machine-generated summary of:

Berggren, Martin; Laudon, Hjalmar; Jonsson, Anders; Jansson, Mats Nutrient Constraints on Metabolism Affect the Temperature Regulation of Aquatic Bacterial Growth Efficiency [123]

Published in: Microbial Ecology (2010)

Link to original: https://doi.org/10.1007/s00248-010-9751-1

If you want to cite the papers, please refer to the original.

For technical reasons we could not place the page where the original quote is coming from.

Research Highlights

Bacterial metabolism is crucial for the cycling of dissolved organic carbon (DOC) in aquatic systems, influencing both food web dynamics and carbon exchange with the atmosphere. Temperature is a key regulator of biological processes, yet its impact on bacterial metabolism in freshwater systems, particularly under varying nutrient conditions, remains poorly understood. Now, a new study by Martin Berggren and colleagues investigates how inorganic nutrient availability influences bacterial metabolic responses to temperature variations in boreal forest streams.

The researchers conducted their study in two boreal forest streams in northern Sweden, representing nutrient-limited and non-nutrient-limited conditions. They measured bacterial production (BP), respiration (BR), and growth efficiency (BGE) across a range of temperatures. Water samples were collected from the streams and incubated with varying concentrations of inorganic nutrients to assess their impact on bacterial metabolism. The study found that BP and BR increased exponentially with temperature, but the response was modulated by nutrient availability. In nutrient-enriched conditions, BR increased more than BP, leading to a decrease in BGE with rising temperatures. Conversely, in nutrient-limited conditions, BGE remained stable across temperatures, suggesting that nutrient limitation can override temperature effects.

The findings highlight the complex interplay between temperature and nutrient availability in regulating bacterial metabolism. The study suggests that future research should explore the role of other environmental factors, such as community structure and viral interactions, which may further influence these dynamics. Understanding these interactions is crucial for predicting how climate change might affect carbon cycling in freshwater ecosystems. The results also imply that nutrient management could be a strategy to mitigate carbon emissions from aquatic systems by influencing bacterial metabolic pathways.

Nutritional Immunity: The Battle for Nutrient Metals at the Host–Pathogen Interface [124]

This is a machine-generated summary of:

Murdoch, Caitlin C.; Skaar, Eric P. Nutritional Immunity: The Battle for Nutrient Metals at the Host–Pathogen Interface [124]

Published in: Nature Reviews Microbiology (2022)
Link to original: https://doi.org/10.1038/s41579-022-00745-6

If you want to cite the papers, please refer to the original.

For technical reasons we could not place the page where the original quote is coming from.

Research Highlights

Trace metals, such as zinc and iron, are crucial for the survival and function of all life forms, playing vital roles in enzymatic reactions, protein structure stabilization, and electron transport. However, both deficiency and excess of these metals can lead to significant health issues in vertebrates. The balance of metal levels is maintained through complex systems that regulate their uptake, storage, and export. Despite the known importance of trace metals in cellular processes, the mechanisms by which vertebrates and bacterial pathogens manage metal homeostasis during infections remain underexplored. Now a new study by Caitlin C. Murdoch and colleagues provides insights into the intricate interactions between hosts and pathogens concerning metal metabolism.

The study explores how vertebrate hosts employ nutritional immunity to limit bacterial access to essential metals like iron (Fe), zinc (Zn), and manganese (Mn) during infections. The authors detail various host strategies for sequestering these metals, such as the use of proteins like transferrin and calprotectin to bind Fe and Zn, respectively. They also discuss how bacterial pathogens have evolved sophisticated mechanisms to counteract these host defenses, including the secretion of siderophores to scavenge Fe and the development of high-affinity metal uptake systems. The research highlights the dynamic interplay at the host–pathogen interface, where both parties continuously adapt their strategies to either sequester or acquire essential nutrients.

The findings underscore the potential for developing novel therapeutic strategies targeting metal homeostasis pathways in bacterial pathogens. Future studies could focus on identifying additional host proteins involved in metal sequestration and exploring bacterial metallophores beyond those that bind iron. Understanding these interactions could lead to innovative treatments that exploit nutritional immunity to combat multidrug-resistant bacterial infections. The study also suggests further investigation into how bacteria prioritize metal allocation during infection could provide deeper insights into pathogen survival strategies within host environments.

Diet Prevents the Expansion of Segmented Filamentous Bacteria and Ileo-Colonic Inflammation in a Model of Crohn's Disease [125]

This is a machine-generated summary of:

Metwaly, Amira; Jovic, Jelena; Waldschmitt, Nadine; Khaloian, Sevana; Heimes, Helena; Häcker, Deborah; Ahmed, Mohamed; Hammoudi, Nassim; Le Bourhis,

Lionel; Mayorgas, Aida; Siebert, Kolja; Basic, Marijana; Schwerd, Tobias; Allez, Matthieu; Panes, Julian; Salas, Azucena; Bleich, André; Zeissig, Sebastian; Schnupf, Pamela; Cominelli, Fabio; Haller, Dirk Diet Prevents the Expansion of Segmented Filamentous Bacteria and Ileo-Colonic Inflammation in a Model of Crohn's Disease [125]

Published in: Microbiome (2023)

Link to original: https://doi.org/10.1186/s40168-023-01508-y

If you want to cite the papers, please refer to the original.

For technical reasons we could not place the page where the original quote is coming from.

Research Highlights

Crohn's disease (CD) is a subtype of inflammatory bowel disease characterized by patchy transmural inflammation in the gastrointestinal tract, particularly affecting the terminal ileum and proximal colon. Despite identifying numerous genetic loci associated with CD, a significant portion of its heritability remains unexplained, suggesting that microbial factors play a crucial role in its etiology. The gut microbiome's perturbations are linked to chronic intestinal inflammation, as seen in CD. Now, a new study by Amira Metwaly and colleagues explores the role of segmented filamentous bacteria (SFB) in driving CD-like inflammation in a mouse model, highlighting the influence of diet on SFB colonization and disease development.

The study utilized Tnf ΔARE mice, a model that mimics human CD-like ileal inflammation. Researchers housed these mice under germ-free conditions and introduced specific-pathogen-free microbiota to observe disease progression. They found that SFB colonization led to severe ileo-colonic inflammation characterized by loss of Paneth and goblet cell functions. To investigate dietary influences, the team fed mice either a fiber-containing chow diet or an exclusive enteral nutrition (EEN)-like purified diet. The purified diet effectively antagonized SFB colonization and prevented disease development, underscoring the critical role of diet in modulating gut microbiota and inflammation. Methods included mono-colonization experiments with SFB and other bacterial strains, microbiome profiling through rRNA gene sequencing, and histopathological assessments to evaluate tissue inflammation.

The findings suggest that SFB plays a pathogenic role in inducing CD-like inflammation in mice, with dietary components significantly influencing SFB colonization and subsequent disease progression. Future study directions could explore the specific dietary components responsible for modulating SFB growth and their potential therapeutic implications for human CD treatment. The study highlights the importance of understanding microbial–diet interactions in managing inflammatory bowel diseases and suggests that dietary interventions like EEN could be effective in controlling pathobiont-driven intestinal inflammation.

A Slam-Dependent Hemophore Contributes to Heme Acquisition in the Bacterial Pathogen **Acinetobacter baumannii** *[126]*

This is a machine-generated summary of:

Bateman, Thomas J.; Shah, Megha; Ho, Timothy Pham; Shin, Hyejin Esther; Pan, Chuxi; Harris, Greg; Fegan, Jamie E.; Islam, Epshita A.; Ahn, Sang Kyun; Hooda, Yogesh; Gray-Owen, Scott D.; Chen, Wangxue; Moraes, Trevor F. A Slam-Dependent Hemophore Contributes to Heme Acquisition in the Bacterial Pathogen *Acinetobacter baumannii* [126]

Published in: Nature Communications (2021)

Link to original: https://doi.org/10.1038/s41467-021-26545-9

If you want to cite the papers, please refer to the original.

For technical reasons we could not place the page where the original quote is coming from.

Research Highlights

Acinetobacter baumannii is a Gram-negative bacterial pathogen responsible for severe infections, including pneumonia and sepsis, and is recognized as a significant public health threat due to its multidrug resistance. The bacterium's ability to persist on surfaces and its genetic adaptability make it particularly challenging to treat with conventional antibiotics. Despite ongoing research efforts, the mechanisms underlying *A. baumannii*'s virulence and pathogenesis remain incompletely understood. Now, a new study by Thomas J. Bateman and colleagues explores the heme acquisition systems of *A. baumannii*, identifying key components that contribute to its virulence.

The study investigates the role of two gene clusters involved in heme utilization, focusing on their contribution to *A. baumannii*'s growth and virulence. The researchers used a combination of genetic knockouts, growth assays, and structural biology techniques to elucidate the function of these clusters. They discovered that the hemO cluster encodes a unique TonB-dependent receptor (HphR) and a secreted heme-binding protein (HphA), which together facilitate heme uptake from human hemoproteins like hemoglobin and albumin. The team demonstrated that HphA acts as a hemophore, binding heme with high affinity and delivering it to HphR for transport across the outer membrane. This system was shown to be critical for *A. baumannii*'s ability to disseminate in the bloodstream and cause systemic infections in mouse models.

The findings highlight the importance of heme acquisition in *A. baumannii*'s pathogenesis and suggest potential targets for therapeutic intervention. The study advances our understanding of bacterial iron acquisition mechanisms, particularly the role of Slam proteins in secreting hemophores like HphA. Future studies could explore whether additional substrates are translocated by Slam proteins in *A. baumannii*, potentially revealing new aspects of its virulence strategy. These insights could inform the development of novel treatments or vaccines aimed at disrupting heme uptake pathways in this formidable pathogen.

Functional Responses of Bacterioplankton Diversity and Metabolism to Experimental Bottom-Up and Top-Down Forcings [127]

This is a machine-generated summary of:

Pradeep Ram, A. S.; Chaibi-Slouma, S.; Keshri, J.; Colombet, J.; Sime-Ngando, T. Functional Responses of Bacterioplankton Diversity and Metabolism to Experimental Bottom-Up and Top-Down Forcings [127]

Published in: Microbial Ecology (2016)
Link to original: https://doi.org/10.1007/s00248-016-0782-0

If you want to cite the papers, please refer to the original.

For technical reasons we could not place the page where the original quote is coming from.

Research Highlights

Bacterioplankton communities are crucial components of aquatic ecosystems, contributing significantly to global genetic diversity and playing vital roles in energy and nutrient cycling. Understanding the factors that control their community composition and activity is essential for elucidating their ecological and biogeochemical functions. While both bottom-up (resource-driven) and top-down (predation-driven) mechanisms are known to influence bacterial diversity, the specific impact of viral infections on bacterial metabolism and community structure remains unclear. Now, a new study by A. S. Pradeep Ram and colleagues investigates the role of viruses in regulating bacterial metabolism through selective lysis of active cells, potentially altering community composition.

The study employed a time-series laboratory microcosm experiment using water samples from Lake Aydat in France to explore the simultaneous effects of predators and resources on bacterial physiological state, metabolism, and community structure. The researchers used high-throughput DNA sequencing to monitor changes in bacterial composition over time under different experimental conditions, including nutrient addition. By filtering water samples to remove larger eukaryotes, they created a grazer-free fraction containing only bacteria and viruses. The experimental setup involved incubating these samples under controlled conditions and analyzing microbial parameters at various time points using flow cytometry, electron microscopy, and metagenomic DNA extraction techniques.

The findings reveal that viral lysis significantly influences the active high nucleic acid content in bacterial cells, impacting carbon metabolism and community structure. The study highlights that nutrient availability enhances bacterial growth efficiency by promoting the proliferation of metabolically active cells, which are then targeted by viruses. This dynamic suggests a "killing the winner" scenario where viruses regulate bacterial communities by suppressing dominant groups, allowing less competitive species to persist. Future studies could further explore how these interactions vary across different aquatic environments and under varying nutrient conditions to better understand their ecological implications.

Nutritional Preferences of Human Gut Bacteria Reveal Their Metabolic Idiosyncrasies [128]

This is a machine-generated summary of:

Tramontano, Melanie; Andrejev, Sergej; Pruteanu, Mihaela; Klünemann, Martina; Kuhn, Michael; Galardini, Marco; Jouhten, Paula; Zelezniak, Aleksej;

Zeller, Georg; Bork, Peer; Typas, Athanasios; Patil, Kiran Raosaheb Nutritional Preferences of Human Gut Bacteria Reveal Their Metabolic Idiosyncrasies [128]

Published in: Nature Microbiology (2018)
Link to original: https://doi.org/10.1038/s41564-018-0123-9

If you want to cite the papers, please refer to the original.

For technical reasons we could not place the page where the original quote is coming from.

Research Highlights

The human gut microbiota plays a crucial role in health and disease, influenced by host factors like the immune system and lifestyle, as well as metabolic interactions among bacterial species. These interactions often involve bacterial metabolites such as vitamins and short-chain fatty acids. However, understanding these complex metabolic interactions is challenging due to the lack of advanced culture-based model systems. Most gut bacteria have been grown in complex media with unknown compositions, limiting mechanistic studies of community functions. Now, a new study by Melanie Tramontano and colleagues addresses this gap by reporting growth profiles of a diverse panel of human gut bacteria across rich and defined media.

The study involved selecting a wide range of phylogenetically and metabolically diverse bacterial strains from healthy human gut microbiota. The researchers developed several defined media by excluding non-defined compounds and adjusting nutrient levels to support the growth of these strains. They evaluated the growth performance of the selected strains under anaerobic conditions across various media, monitoring optical density over time to assess growth. The study found that most strains grew in at least one defined medium, with some showing better growth in defined media than in rich ones. The researchers also explored the effects of short-chain fatty acids (SCFAs) and amino acids on bacterial growth, finding that SCFAs inhibited some species while promoting others.

The findings provide valuable insights into the nutritional preferences and biosynthetic capabilities of gut bacteria, highlighting the potential for defined media to improve metabolic modeling and understanding of gut microbiome ecology. Future studies could explore the role of mucin-utilizing bacteria in health and disease, as well as refine metabolic models using this new resource. This work represents a significant step toward understanding the structure and function of the human gut microbiome, offering a foundation for further research into microbial interactions and their implications for host health.

5.3 Conclusion

The studies presented in this chapter collectively feature the central role of nutrient acquisition in bacterial physiology, ecology, and pathogenesis. Far from being passive consumers of environmental resources, bacteria exhibit highly adaptive and often cooperative strategies to secure essential nutrients under diverse and often hostile conditions. These strategies not only support survival and growth but also shape microbial interactions, influence host responses, and determine disease outcomes. As such, nutrient acquisition emerges as a critical axis of microbial life, with profound implications for health, disease, and biotechnology.

One of the most compelling themes in this chapter is the cooperative behavior that emerges under nutrient stress. The interaction between *Desulfovibrio vulgaris* and *Clostridium acetobutylicum* illustrates how nutrient limitation can drive physical and metabolic coupling between species. This exchange of cytoplasmic material enables survival and enhances metabolic output, such as hydrogen production. These findings challenge the traditional view of microbial competition and highlight the importance of cooperation in microbial consortia, particularly in nutrient-limited environments. Such insights could inform the design of synthetic microbial communities for industrial or environmental applications.

The regulatory role of specific nutrients, such as L-glutamine, in shaping microbial metabolism is another key insight. Glutamine's ability to modulate amino acid utilization in intestinal bacteria suggests that dietary components can influence microbial metabolic networks in a species-specific manner. This has important implications for gut health, as nutrient availability can affect microbial composition, metabolic output, and interactions with the host immune system. These findings support the growing interest in dietary interventions to modulate the gut microbiome and improve health outcomes.

Environmental conditions, including oxygen levels and nutrient richness, also play a pivotal role in regulating bacterial virulence traits. In *Mycobacterium avium* subsp. *hominissuis*, eutrophic, and hypoxic conditions promote biofilm-like pellicle formation, a key factor in persistence and resistance. The role of glycopeptidolipids (GPLs) in this process highlights the molecular mechanisms by which environmental cues influence pathogenic behavior. Understanding these mechanisms could lead to novel strategies for disrupting biofilm formation and controlling infections caused by nontuberculous mycobacteria.

At the cellular level, the synergistic effects of nutrient and salt depletion on glucose metabolism in *Escherichia coli* reveal a previously unrecognized level of metabolic regulation. These changes occur independently of transcriptional responses and are instead mediated by translational and posttranslational mechanisms. This finding expands our understanding of bacterial adaptability and suggests new targets for manipulating metabolic pathways in industrial or clinical settings.

The interplay between diet and immunity further emphasizes the systemic impact of nutrient availability. Dietary components that interact with nuclear receptors in immune cells can modulate T cell differentiation and mucosal immunity, influencing susceptibility to inflammatory diseases. These insights highlight the potential of personalized nutrition to support immune homeostasis and prevent disease.

The evolutionary dynamics of microbial cooperation are also shaped by nutrient availability. In *Pseudomonas aeruginosa*, the fitness cost of siderophore production varies depending on which nutrient is limiting, influencing the balance between cooperative and cheating strains. This finding provides a physiological basis for understanding the stability of cooperative behaviors in microbial populations and suggests that environmental context is critical in shaping social evolution.

In aquatic ecosystems, nutrient constraints modulate the temperature sensitivity of bacterial growth efficiency, with implications for carbon cycling and climate change. Nutrient-limited conditions buffer the effects of rising temperatures, suggesting that nutrient management could be a strategy for mitigating microbial contributions to greenhouse gas emissions.

Finally, the concept of nutritional immunity illustrates the competitive dynamics at the host–pathogen interface. Hosts sequester essential metals to limit bacterial growth, while pathogens evolve high-affinity uptake systems to overcome these defenses. Understanding these interactions opens new avenues for antimicrobial development, particularly in the face of rising antibiotic resistance.

In conclusion, bacterial nutrient acquisition is a dynamic and integrative process that influences microbial behavior, community structure, and host interactions. The studies in this chapter reveal the complexity and adaptability of bacterial nutrient strategies and highlight their importance in both health and disease. As we deepen our understanding of these processes, new opportunities will emerge for manipulating microbial ecosystems, developing targeted therapies, and promoting human and environmental health.

Bibliography

1. Bryant, M. P., Campbell, L. L., Reddy, C. A. & Crabill, M. R. Growth of desulfovibrio in lactate or ethanol media low in sulfate in association with H2-utilizing methanogenic bacteria. *Appl. Environ. Microbiol. 33*, 1162–1169 (1977).
2. Traore, A. S., Fardeau, M. L., Hatchikian, C. E., Le Gall, J. & Belaich, J. P. Energetics of growth of a defined mixed culture of *Desulfovibrio vulgaris* and *Methanosarcina barkeri*: interspecies hydrogen transfer in batch and continuous cultures. *Appl. Environ. Microbiol. 46*, 1152–1156 (1983).
3. Walker, C. B. et al. The electron transfer system of syntrophically grown *Desulfovibrio vulgaris*. *J. Bacteriol. 191*, 5793–5801 (2009).
4. Collet, C., Gaudard, O., Péringer, P. & Schwitzguébel, J. P. Acetate production from lactose by *Clostridium thermolacticum* and hydrogen-scavenging microorganisms in continuous culture–effect of hydrogen partial pressure. *J. Biotechnol. 118*, 328–338 (2005).
5. Collet, C., Schwitzguébel, J. P. & Péringer, P. Improvement of acetate production from lactose by growing *Clostridium thermolacticum* in mixed batch culture. *J. Appl. Microbiol. 95*, 824–831 (2003).

6. Muyzer, G. & Stams, A. J. The ecology and biotechnology of sulphate-reducing bacteria. *Nat. Rev. Microbiol. 6*, 441–454 (2008).
7. Dai ZL, Zhang J, Wu G et al (2010) Utilization of amino acids by bacteria from the pig small intestine. Amino Acids 39:1201–1215
8. Dai ZL, Li XL, Xi PB et al (2011a) Metabolism of select amino acids in bacteria from the pig small intestine. Amino Acids. https://doi.org/10.1007/s00726-011-0846-x
9. Metges CC, Petzke KJ (2005) Utilization of essential amino acids synthesized in the intestinal microbiota of monogastric mammals. Br J Nutr 94:621–622
10. Forchhammer K (2007) Glutamine signalling in bacteria. Front Biosci 12:358–370
11. Dai ZL, Li XL, Xi PB et al (2011b) Regulatory role for *l*-arginine in the utilization of amino acids by pig small-intestinal bacteria. Amino Acids. https://doi.org/10.1007/s00726-011-1067-z
12. Reeds PJ, Burrin DG (2001) Glutamine and the bowel. J Nutr 131(9 Suppl):2505S–2508S
13. Wu G, Borbolla AG, Knabe DA (1994) The uptake of glutamine and release of arginine, citrulline and proline by the small intestine of developing pigs. J Nutr 124:2437–2444
14. Wu G, Meier SA, Knabe DA (1996) Dietary glutamine supplementation prevents jejunal atrophy in weaned pigs. J Nutr 126:2578–2584
15. Wu G (1998) Intestinal mucosal amino acid catabolism. J Nutr 128:1249–1252
16. Wang JJ, Chen LX, Li P et al (2008) Gene expression is altered in piglet small intestine by weaning and dietary glutamine supplementation. J Nutr 138:1025–1032
17. Wu G, Bazer FW, Davis TA et al (2007b) Important roles for the arginine family of amino acids in swine nutrition and production. Livest Sci 112:8–22
18. Wu G, Bazer FW, Davis TA et al (2009) Arginine metabolism and nutrition in growth, health and disease. Amino Acids 37:153–168
19. Satterfield MC, Dunlap KA, Keisler DH et al (2011) Arginine nutrition and fetal brown adipose tissue development in nutrient-restricted sheep. Amino Acids. https://doi.org/10.1007/s00726-011-1168-8
20. Satterfield MC, Dunlap KA, Keisler DH et al (2012) Arginine nutrition and fetal brown adipose tissue development in diet-induced obese sheep. Amino Acids. https://doi.org/10.1007/s00726-012-1235-9
21. Falkinham, J. O. 3rd, Norton, C. D. & LeChevallier, M. W. Factors influencing numbers of *Mycobacterium avium*, *Mycobacterium intracellulare*, and other mycobacteria in drinking water distribution systems. *Appl Environ Microbiol. 67*, 1225–1231 (2001).
22. Falkinham, J. O. 3rd. Environmental sources of *Mycobacterium avium* linked to routes of exposure. (eds Pedley, S. et al.) Ch. 3, 26–38 (IWA Publishing, 2004).
23. Williams, M. M. et al. Structural analysis of biofilm formation by rapidly and slowly growing nontuberculous mycobacteria. *Appl Environ Microbiol. 75*, 2091–2098 (2009).
24. Ojha, A. et al. GroEL1: a dedicated chaperone involved in mycolic acid biosynthesis during biofilm formation in mycobacteria. *Cell. 123*, 861–873 (2005).
25. Ojha, A. K. et al. Growth of *Mycobacterium tuberculosis* biofilms containing free mycolic acids and harbouring drug-tolerant bacteria. *Mol Microbiol. 69*, 164–174 (2008).
26. Sambandan, D. et al. Keto-mycolic acid-dependent pellicle formation confers tolerance to drug-sensitive *Mycobacterium tuberculosis* . *mBio. 4*, e00222–00213 (2013).
27. Kerns, P. W., Ackhart, D. F., Basaraba, R. J., Leid, J. G. & Shirtliff, M. E. *Mycobacterium tuberculosis* pellicles express unique proteins recognized by the host humoral response. *Pathog Dis. 70*, 347–358 (2014).
28. Pang, J. M. et al. The polyketide Pks1 contributes to biofilm formation in *Mycobacterium tuberculosis*. *J Bacteriol. 194*, 715–721 (2012).
29. Wayne, L. G. & Hayes, L. G. An *in vitro* model for sequential study of shiftdown of *Mycobacterium tuberculosis* through two stages of nonreplicating persistence.. *Infect Immun. 64*, 2062–2069 (1996).
30. Via, L. E. et al. Tuberculous granulomas are hypoxic in guinea pigs, rabbits, and nonhuman primates. *Infect Immun. 76*, 2333–2340 (2008).

31. Kumar, A., Toledo, J. C., Patel, R. P., Lancaster, J. R. Jr. & Steyn, A. J. *Mycobacterium tuberculosis* DosS is a redox sensor and DosT is a hypoxia sensor. *Proc Natl Acad Sci USA 104*, 11568–11573 (2007).
32. Lee, J. M. et al. O2- and NO-sensing mechanism through the DevSR two-component system in *Mycobacterium smegmatis* . *J Bacteriol. 190*, 6795–6804 (2008).
33. Mayuri, Bagchi, G., Das, T. K. & Tyagi, J. S. Molecular analysis of the dormancy response in *Mycobacterium smegmatis*: expression analysis of genes encoding the DevR-DevS two-component system, Rv3134c and chaperone alpha-crystallin homologues. *FEMS Microbiol Lett. 211*, 231–237 (2002).
34. Nishiuchi, Y. et al. The recovery of *Mycobacterium avium-intracellulare* complex (MAC) from the residential bathrooms of patients with pulmonary MAC. *Clin Infect Dis. 45*, 347–351 (2007).
35. Nishiuchi, Y. et al. *Mycobacterium avium* complex organisms predominantly colonize in the bathtub inlets of patients' bathrooms. *Jpn J Infect Dis. 62*, 182–186 (2009).
36. Henkle, E., Hedberg, K., Schafer, S., Novosad, S. & Winthrop, K. L. Population-based incidence of pulmonary nontuberculous mycobacterial disease in Oregon 2007 to 2012. *Ann Am Thorac Soc. 12*, 642–647 (2015).
37. Namkoong, H. et al. Epidemiology of pulmonary nontuberculous mycobacterial disease in Japan. *Emerg Infect Dis. 22*, 1116–1117 (2016).
38. Schaefer, U., Boos, W., Takors, R. & Weuster-Botz, D. Automated sampling device for monitoring intracellular metabolite dynamics. *Anal. Biochem. 270*, 88–96 (1999).
39. Chubukov, V. et al. Engineering glucose metabolism of escherichia coli under nitrogen starvation. *npj Syst. Biol. Appl. 3*, 1–7 (2017).
40. Nestorovich, E. M., Danelon, C., Winterhalter, M. & Bezrukov, S. M. Designed to penetrate: Time-resolved interaction of single antibiotic molecules with bacterial pores. *Proc. Natl Acad. Sci. U. S. A. 99*, 9789–9794 (2002).
41. Cama, J. et al. Single-cell microfluidics facilitates the rapid quantification of antibiotic accumulation in Gram-negative bacteria. *Lab Chip 20*, 2765–2775 (2020).
42. Łapińska, U. et al. Fast bacterial growth reduces antibiotic accumulation and efficacy. *Biorxiv* https://doi.org/10.1101/2021.10.18.464851v1 (2021).
43. Braissant, O., Astasov-Frauenhoffer, M., Waltimo, T. & Bonkat, G. A Review of Methods to Determine Viability, Vitality, and Metabolic Rates in Microbiology. *Front. Microbiol. 11*, 1–25 (2020).
44. Nikaido H (2003) Molecular basis of bacterial outer membrane permeability revisited. Microbiol Mol Biol Rev 67:593–656. https://doi.org/10.1128/mmbr.67.4.593-656.2003.
45. Pratt, L. A., Hsing, W., Gibson, K. E. & Silhavy, T. J. From acids to osmZ: Multiple factors influence synthesis of the OmpF and OmpC porins in Escherichia coli. *Mol. Microbiol. 20*, 911–917 (1996).
46. Chubiz, L. M. & Rao, C. V. Role of the mar-sox-rob regulon in regulating outer membrane porin expression. *J. Bacteriol. 193*, 2252–2260 (2011).
47. Natarajan, A. & Srienc, F. Dynamics of glucose uptake by single Escherichia coli cells. *Metab. Eng. 1*, 320–333 (1999).
48. Natarajan, A. & Srienc, F. Glucose uptake rates of single E. coli cells grown in glucose-limited chemostat cultures. *J. Methods Microbiol. J. Microbiol. Methods 42*, 87–96 (2000).
49. Lendenmann, U. & Egli, T. Is Escherichia coli growing in glucose-limited chemostat culture able to utilize other sugars without lag? *Microbiology 141*, 71–78 (1995).
50. Herbert, D. & Kornberg, H. L. Glucose transport as rate-limiting step in the growth of Escherichia coli on Glucose. *Biochem. J. 156*, 477–480 (1976).
51. Stiers, K. M., Muenks, A. G. & Beamer, L. J. Biology, mechanism, and structure of enzymes in the α-D-phosphohexomutase superfamily. *Adv. Protein Chem. Struct. Biol. 176*, 139–148 (2017).
52. Bren, A. et al. Glucose becomes one of the worst carbon sources for E.coli on poor nitrogen sources due to suboptimal levels of cAMP. *Sci. Rep. 6*, 1–10 (2016).

53. Jõers, A. & Tenson, T. Growth resumption from stationary phase reveals memory in Escherichia coli cultures. *Sci. Rep. 6*, 1–11 (2016).
54. Brockhurst, M.A., Buckling, A., Racey, D. & Gardner, A. Resource supply and the evolution of public-goods cooperation in bacteria. *BMC Biol. 6*, 20 (2008).
55. Buckling, A. et al. Siderophore-mediated cooperation and virulence in *Pseudomonas aeruginosa. FEMS Microbiol. Ecol. 62*, 135–141 (2007).
56. Scholz, R. L. & Greenberg, E. P. Sociality in *Escherichia coli*: enterochelin is a private good at low cell density and can be shared at high cell density. *J. Bacteriol. 197*, 2122–2128 (2015).
57. Yeterian, E. et al. Synthesis of the siderophore pyoverdine in *Pseudomonas aeruginosa* involves a periplasmic maturation. *Amino acids 38*, 1447–1459 (2010).
58. Pomeroy LR, Wiebe WJ (2001) Temperature and substrates as interactive limiting factors for marine heterotrophic bacteria. Aquat Microb Ecol 23:187–204
59. Rivkin RB, Legendre L (2001) Biogenic carbon cycling in the upper ocean: effects of microbial respiration. Science 291:2398–2400
60. Lopez-Urrutia A, Moran XAG (2007) Resource limitation of bacterial production distorts the temperature dependence of oceanic carbon cycling. Ecology 88:817–822
61. Berggren M, Laudon H, Jansson M (2007) Landscape regulation of bacterial growth efficiency in boreal freshwaters. Global Biogeochem Cycles 21:GB4002
62. Reinthaler T, van Aken H, Veth C, Aristegui J, Robinson C, Williams P, Lebaron P, Herndl GJ (2006) Prokaryotic respiration and production in the meso- and bathypelagic realm of the eastern and western North Atlantic basin. Limnol Oceanogr 51:1262–1273
63. Pradeep Ram AS, Nishimura Y, Tomaru Y, Nagasaki K, Nagata T (2010) Seasonal variation in viral-induced mortality of bacterioplankton in the water column of a large mesotrophic lake (Lake Biwa, Japan). Aquat Microb Ecol 58:249–259
64. Hood, M. I. & Skaar, E. P. Nutritional immunity: transition metals at the pathogen–host interface. *Nat. Rev. Microbiol. 10*, 525–537 (2012).
65. Singh, V. et al. Microbiota-inducible innate immune, siderophore binding protein lipocalin 2 is critical for intestinal homeostasis. *Cell Mol. Gastroenterol. Hepatol. 2*, 482–498.e6 (2016).
66. Choby, J. E. & Skaar, E. P. Heme synthesis and acquisition in bacterial pathogens. *J. Mol. Biol. 428*, 3408–3428 (2016).
67. Sheldon, J. R., Laakso, H. A. & Heinrichs, D. E. Iron acquisition strategies of bacterial pathogens. *Microbiol. Spectr.* https://doi.org/10.1128/microbiolspec.VMBF-0010-2015 (2016).
68. Perez, N., Johnson, R., Sen, B. & Ramakrishnan, G. Two parallel pathways for ferric and ferrous iron acquisition support growth and virulence of the intracellular pathogen *Francisella tularensis* Schu S4. *Microbiologyopen 5*, 453–468 (2016).
69. Monteith, A. J. & Skaar, E. P. The impact of metal availability on immune function during infection. *Trends Endocrinol. Metab. 32*, 916–928 (2021).
70. Damo, S. M. et al. Molecular basis for manganese sequestration by calprotectin and roles in the innate immune response to invading bacterial pathogens. *Proc. Natl Acad. Sci. USA 110*, 3841–3846 (2013).
71. Nakashige, T. G., Zygiel, E. M., Drennan, C. L. & Nolan, E. M. Nickel sequestration by the host-defense protein human calprotectin. *J. Am. Chem. Soc. 139*, 8828–8836 (2017).
72. Cerasi, M. et al. The ZupT transporter plays an important role in zinc homeostasis and contributes to *Salmonella enterica* virulence. *Metallomics 6*, 845–853 (2014).
73. Grass, G., Wong, M. D., Rosen, B. P., Smith, R. L. & Rensing, C. ZupT is a Zn(II) uptake system in *Escherichia coli. J. Bacteriol. 184*, 864–866 (2002).
74. Sabri, M., Houle, S. & Dozois, C. M. Roles of the extraintestinal pathogenic *Escherichia coli* ZnuACB and ZupT zinc transporters during urinary tract infection. *Infect. Immun. 77*, 1155–1164 (2009).
75. Corbett, D. et al. Two zinc uptake systems contribute to the full virulence of *Listeria monocytogenes* during growth in vitro and in vivo. *Infect. Immun. 80*, 14–21 (2012).

76. Rosadini, C. V., Gawronski, J. D., Raimunda, D., Arguello, J. M. & Akerley, B. J. A novel zinc binding system, ZevAB, is critical for survival of nontypeable *Haemophilus influenzae* in a murine lung infection model. *Infect. Immun. 79*, 3366–3376 (2011).
77. Jean, S., Juneau, R. A., Criss, A. K. & Cornelissen, C. N. *Neisseria gonorrhoeae* evades calprotectin-mediated nutritional immunity and survives neutrophil extracellular traps by production of TdfH. *Infect. Immun. 84*, 2982–2994 (2016).
78. Maurakis, S. et al. The novel interaction between *Neisseria gonorrhoeae* TdfJ and human S100A7 allows gonococci to subvert host zinc restriction. *PLoS Pathog. 15*, e1007937 (2019).
79. Stork, M. et al. Zinc piracy as a mechanism of *Neisseria meningitidis* for evasion of nutritional immunity. *PLoS Pathog. 9*, e1003733 (2013).
80. Ladomersky, E. & Petris, M. J. Copper tolerance and virulence in bacteria. *Metallomics 7*, 957–964 (2015).
81. Zhanel, G. G. et al. Cefiderocol: a siderophore cephalosporin with activity against carbapenem-resistant and multidrug-resistant Gram-negative bacilli. *Drugs 79*, 271–289 (2019).
82. Schaubeck M, et al. Dysbiotic gut microbiota causes transmissible Crohn's disease-like ileitis independent of failure in antimicrobial defence. Gut. 2016;65:225–37.
83. Metwaly A, et al. Integrated microbiota and metabolite profiles link Crohn's disease to sulfur metabolism. Nat Commun. 2020;11:4322.
84. Corraliza AM, et al. Differences in peripheral and tissue immune cell populations following haematopoietic stem cell transplantation in Crohn's disease patients. J Crohns Colitis. 2019;13:634–47.
85. de Léséleuc, L., Harris, G., KuoLee, R., Xu, H. H. & Chen, W. Serum resistance, gallium nitrate tolerance and extrapulmonary dissemination are linked to heme consumption in a bacteremic strain of *Acinetobacter baumannii*. *Int. J. Med. Microbiol. 304*, 360–369 (2014).
86. Hooda, Y. et al. Slam is an outer membrane protein that is required for the surface display of lipidated virulence factors in *Neisseria*. *Nat. Microbiol. 1*, 16009 (2016).
87. Giardina, B. J., Shahzad, S., Huang, W. & Wilks, A. Heme uptake and utilization by hypervirulent *Acinetobacter baumannii* LAC-4 is dependent on a canonical heme oxygenase (abHemO). *Arch. Biochem. Biophysics 672*, 108066 (2019).
88. Latham, R. D. et al. A heme-binding protein produced by *Haemophilus haemolyticus* inhibits non-typeable *Haemophilus influenzae*. *Mol. Microbiol. 113*, 381–398 (2020).
89. Fanelli, A. R. Studies on the structure of hemoglobin I. Physicochemical properties of human globin. *BBA - Biochimica et. Biophysica Acta 30*, 608–615 (1958).
90. Espinas, N. A., Kobayashi, K., Takahashi, S., Mochizuki, N. & Masuda, T. Evaluation of unbound free heme in plant cells by differential acetone extraction. *Plant Cell Physiol. 53*, 1344–1354 (2012).
91. Runyen-Janecky, L. J. Role and regulation of heme iron acquisition in gram-negative pathogens. *Front. Cell Infect. Microbiol. 3*, 55 (2013).
92. Wójtowicz, H. et al. Unique structure and stability of HmuY, a novel heme-binding protein of *Porphyromonas gingivalis*. *PLOS Pathog. 5*, e1000419 (2009).
93. Smalley, J. W. et al. HmuY Haemophore and Gingipain Proteases Constitute a Unique Syntrophic System of Haem Acquisition by *Porphyromonas gingivalis*. *PLoS ONE 6*, e17182 (2011).
94. Bielecki, M. et al. Tannerella forsythia Tfo belongs to *Porphyromonas gingivalis* HmuY-like family of proteins but differs in heme-binding properties. *Biosci. Rep. 38*, https://doi.org/10.1042/BSR20181325 (2018).
95. Krieg, S. et al. Heme uptake across the outer membrane as revealed by crystal structures of the receptor-hemophore complex. *Proc. Natl Acad. Sci. USA 106*, 1045–1050 (2009).
96. Bouvier T, del Giorgio PA (2007) Key role of selective viral-induced mortality in determining marine bacterial community composition. Environ Microbiol 9:287–297
97. Graham EB et al (2016) Microbes as engines of ecosystem function: when does community structure enhance prediction of ecosystem processes? Front Microbiol 7:214

98. Jardillier L, Boucher D, Personnic S, Jacquet S, Thénot A, Sargos D, Amblard C, Debroas D (2005) Relative importance of nutrients and mortality factors on prokaryotic community composition in two lakes of different trophic status: microcosm experiments. FEMS Microb Ecol 53:429–443
99. Xu J, Jing H, Sun M, Harrison PJ, Liu H (2013) Regulation of bacterial metabolic activity by dissolved organic carbon and viruses. J Geophys Res: Biogeosciences 118:1573–1583
100. Pradeep Ram AS, Colombet J, Perriere F, Thouvenot A, Sime-Ngando T (2015) Viral and grazer regulation of prokaryotic growth efficiency in temperate freshwater pelagic environments. FEMS Microb Ecol 91:1–12
101. Maurice CF, Bouvier T, Comte J, Guillemette F, del Giorgio PA (2010) Seasonal variations of phage life strategies and bacterial physiological states in three northern temperate lakes. Environ Microbiol 12:628–641
102. Pradeep Ram AS, Sime-Ngando T (2014) Distinctive patterns in prokaryotic community composition in response to viral lysis and flagellate grazing in freshwater microcosms. Freshw Biol 59:1945–1955
103. Neidhardt, F. C., Bloch, P. L. & Smith, D. F. Culture medium for enterobacteria. *J. Bacteriol. 119*, 736–747 (1974).
104. Goodman, A. L. et al. Identifying genetic determinants needed to establish a human gut symbiont in its habitat. *Cell Host Microbe 6*, 279–289 (2009).
105. Sebald, M. & Costilow, R. N. Minimal growth requirements for *Clostridium perfringens* and isolation of auxotrophic mutants. *Appl. Microbiol. 29*, 1–6 (1975).
106. Lopes, J. N. & Cruz, F. S. Chemically defined media for growing anaerobic bacteria of the genus *Veillonella. Antonie Van. Leeuwenhoek 42*, 411–420 (1976).
107. Magnúsdóttir, S. et al. Generation of genome-scale metabolic reconstructions for 773 members of the human gut microbiota. *Nat. Biotechnol. 35*, 81–89 (2017).
108. Larsbrink, J. et al. A discrete genetic locus confers xyloglucan metabolism in select human gut Bacteroidetes. *Nature 506*, 498–502 (2014).
109. Thiele, I. & Palsson, B. Ø. A protocol for generating a high-quality genome-scale metabolic reconstruction. *Nat. Protoc. 5*, 93–121 (2010).
110. Shoaie, S. et al. Quantifying diet-induced metabolic changes of the human gut microbiome. *Cell Metab. 22*, 320–331 (2015).
111. Goodman, A. L. et al. Extensive personal human gut microbiota culture collections characterized and manipulated in gnotobiotic mice. *Proc. Natl Acad. Sci. USA 108*, 6252–6257 (2011).
112. Rettedal, E. A., Gumpert, H. & Sommer, M. O. A. Cultivation-based multiplex phenotyping of human gut microbiota allows targeted recovery of previously uncultured bacteria. *Nat. Commun. 5*, 4714 (2014).
113. Browne, H. P. et al. Culturing of 'unculturable' human microbiota reveals novel taxa and extensive sporulation. *Nature 533*, 543–546 (2016).
114. Lau, J. T. et al. Capturing the diversity of the human gut microbiota through culture-enriched molecular profiling. *Genome Med. 8*, 72 (2016).
115. Maier, L. et al. Extensive impact of non-antibiotic drugs on human gut bacteria. *Nature* https://doi.org/10.1038/nature25979 (2018).
116. Duncan, S. H., Hold, G. L., Harmsen, H. J. M., Stewart, C. S. & Flint, H. J. Growth requirements and fermentation products of *Fusobacterium prausnitzii*, and a proposal to reclassify it as *Faecalibacterium prausnitzii* gen. nov., comb. nov. *Int. J. Syst. Evol. Microbiol. 52*, 2141–2146 (2002).
117. Benomar, Saida; Ranava, David; Cárdenas, María Luz; Trably, Eric; Rafrafi, Yan; Ducret, Adrien; Hamelin, Jérôme; Lojou, Elisabeth; Steyer, Jean-Philippe; Giudici-Orticoni, Marie-Thérèse Nutritional stress induces exchange of cell material and energetic coupling between bacterial species. Nature Communications (2015). https://doi.org/10.1038/ncomms7283
118. Dai, Zhao-Lai; Li, Xi-Long; Xi, Peng-Bin; Zhang, Jing; Wu, Guoyao; Zhu, Wei-Yun l-Glutamine regulates amino acid utilization by intestinal bacteria. Amino Acids (2012). https://doi.org/10.1007/s00726-012-1264-4

119. Totani, Takahiro; Nishiuchi, Yukiko; Tateishi, Yoshitaka; Yoshida, Yutaka; Kitanaka, Hiromi; Niki, Mamiko; Kaneko, Yukihiro; Matsumoto, Sohkichi. Effects of nutritional and ambient oxygen condition on biofilm formation in Mycobacterium avium subsp. hominissuis via altered glycolipid expression. Scientific Reports (2017). https://doi.org/10.1038/srep41775
120. Glover, Georgina; Voliotis, Margaritis; Łapińska, Urszula; Invergo, Brandon M.; Soanes, Darren; O'Neill, Paul; Moore, Karen; Nikolic, Nela; Petrov, Peter G.; Milner, David S.; Roy, Sumita; Heesom, Kate; Richards, Thomas A.; Tsaneva-Atanasova, Krasimira; Pagliara, Stefano. Nutrient and salt depletion synergistically boosts glucose metabolism in individual Escherichia coli cells. Communications Biology (2022). https://doi.org/10.1038/s42003-022-03336-6
121. Veldhoen, Marc; Brucklacher-Waldert, Verena. Dietary influences on intestinal immunity. Nature Reviews Immunology (2012). https://doi.org/10.1038/nri3299
122. Sexton, D. Joseph; Schuster, Martin. Nutrient limitation determines the fitness of cheaters in bacterial siderophore cooperation. Nature Communications (2017). https://doi.org/10.1038/s41467-017-00222-2
123. Berggren, Martin; Laudon, Hjalmar; Jonsson, Anders; Jansson, Mats. Nutrient Constraints on Metabolism Affect the Temperature Regulation of Aquatic Bacterial Growth Efficiency. Microbial Ecology (2010). https://doi.org/10.1007/s00248-010-9751-1
124. Murdoch, Caitlin C.; Skaar, Eric P. Nutritional immunity: the battle for nutrient metals at the host–pathogen interface. Nature Reviews Microbiology (2022). https://doi.org/10.1038/s41579-022-00745-6
125. Metwaly, Amira; Jovic, Jelena; Waldschmitt, Nadine; Khaloian, Sevana; Heimes, Helena; Häcker, Deborah; Ahmed, Mohamed; Hammoudi, Nassim; Le Bourhis, Lionel; Mayorgas, Aida; Siebert, Kolja; Basic, Marijana; Schwerd, Tobias; Allez, Matthieu; Panes, Julian; Salas, Azucena; Bleich, André; Zeissig, Sebastian; Schnupf, Pamela; Cominelli, Fabio; Haller, Dirk Diet prevents the expansion of segmented filamentous bacteria and ileo-colonic inflammation in a model of Crohn's disease. Microbiome (2023). https://doi.org/10.1186/s40168-023-01508-y
126. Bateman, Thomas J.; Shah, Megha; Ho, Timothy Pham; Shin, Hyejin Esther; Pan, Chuxi; Harris, Greg; Fegan, Jamie E.; Islam, Epshita A.; Ahn, Sang Kyun; Hooda, Yogesh; Gray-Owen, Scott D.; Chen, Wangxue; Moraes, Trevor F. A Slam-dependent hemophore contributes to heme acquisition in the bacterial pathogen Acinetobacter baumannii. Nature Communications (2021). https://doi.org/10.1038/s41467-021-26545-9
127. Pradeep Ram, A. S.; Chaibi-Slouma, S.; Keshri, J.; Colombet, J.; Sime-Ngando, T. Functional Responses of Bacterioplankton Diversity and Metabolism to Experimental Bottom-Up and Top-Down Forcings. Microbial Ecology (2016). https://doi.org/10.1007/s00248-016-0782-0
128. Tramontano, Melanie; Andrejev, Sergej; Pruteanu, Mihaela; Klünemann, Martina; Kuhn, Michael; Galardini, Marco; Jouhten, Paula; Zelezniak, Aleksej; Zeller, Georg; Bork, Peer; Typas, Athanasios; Patil, Kiran Raosaheb. Nutritional preferences of human gut bacteria reveal their metabolic idiosyncrasies. Nature Microbiology (2018). https://doi.org/10.1038/s41564-018-0123-9

Chapter 6
Metabolic Reprogramming of Bacterial and Host Pathways and Their Role in Infections

Vijay Soni and Aditya Upadhyay

6.1 Introduction by the Editor

Metabolism is not merely a background process in cellular life—it is a dynamic and responsive system that shapes the trajectory of infection, immunity, and disease. In recent years, the concept of metabolic reprogramming has emerged as a central theme in understanding host–pathogen interactions. This chapter explores how both bacterial pathogens and host cells undergo metabolic shifts during infection, revealing a complex interplay that influences virulence, immune evasion, and therapeutic outcomes.

One of the most compelling examples of pathogen-driven metabolic reprogramming is seen in *Salmonella Typhimurium*, which manipulates macrophage metabolism to promote its intracellular replication. By deploying the T3SS effector SopE2, *Salmonella* induces a Warburg-like effect enhancing glycolysis and suppressing oxidative phosphorylation. This shift not only supports bacterial growth by increasing the availability of glycolytic intermediates but also dampens host immune responses. The pathogen's ability to sense and respond to host metabolic cues via the CreBC system underscores the sophistication of bacterial adaptation strategies.

Similarly, *Escherichia coli* demonstrates remarkable metabolic flexibility when grown on different carbon sources. Transcriptomic analyses reveal that pyruvate versus glucose as a sole carbon source leads to distinct gene expression profiles, particularly in small RNAs (sRNAs) that regulate central carbon metabolism. These findings highlight the role of sRNAs in fine-tuning bacterial responses to environmental changes and suggest potential targets for metabolic engineering.

V. Soni (✉)
Department of Medicine, Weill Cornell Medicine, New York, NY, USA
e-mail: vis2032@med.cornell.edu

A. Upadhyay
Hospital for Special Surgery, New York, NY, USA

V. Soni, A. Upadhyay (eds.), *Metabolism of Bacterial Infection*,
https://doi.org/10.1007/978-3-031-46895-7_6

The gut microbiome's influence on host metabolism and disease is another major theme. Bacterial metabolites such as short-chain fatty acids, tryptophan derivatives, and lipid components interact with host receptors to modulate inflammation, immunity, and even tumor suppression. In the context of colon cancer, these metabolites act as molecular messengers, linking microbial activity to host cellular pathways. This chemical dialogue between microbes and host cells is crucial for maintaining intestinal homeostasis and preventing disease.

Viral infections also reprogram host metabolism, as demonstrated in studies of bacteriophage-infected *E. coli*. Using systems biology and flux balance analysis, researchers show that viral replication redirects host metabolic fluxes toward amino acid biosynthesis and the pentose phosphate pathway, while suppressing the TCA cycle. These changes mirror metabolic shifts seen in cancer cells, suggesting shared strategies among diverse pathogens to hijack host resources.

Host–microbe co-metabolism further illustrates the integration of microbial and host metabolic pathways. For example, microbial metabolites processed by host enzymes such as medium-chain acyl-CoA dehydrogenase (MCAD) generate circulating compounds like hippuric acid. These metabolites reflect the combined metabolic activities of both host and microbiota, with implications for personalized nutrition and disease risk.

Biofilm formation, a key factor in chronic infections, is also driven by metabolic reprogramming. In uropathogenic *E. coli*, metabolomic profiling reveals distinct shifts in amino acids, carbohydrates, and polyamines during biofilm development. These changes support the structural integrity and antibiotic resistance of biofilms, offering potential targets for disrupting persistent infections.

Staphylococcus aureus exemplifies metabolic adaptation in host environments. Genomic and metabolomic analyses reveal how this pathogen modulates host immune responses and biofilm formation through metabolic shifts. Small colony variants (SCVs) of *S. aureus* further manipulate host cell glycolysis and induce necroptosis, facilitating bacterial persistence without relying on toxins.

Host responses to infection also involve metabolic reprogramming. Extracellular vesicles (EVs) released during *Salmonella* infection act as iron catchers, sequestering iron to limit bacterial access. This humoral mechanism complements cellular nutritional immunity and highlights the systemic nature of host defense strategies.

Finally, *Chlamydia trachomatis* reprograms host glutamine metabolism to support its developmental cycle and peptidoglycan synthesis. By hijacking host glutaminolysis, *Chlamydia* ensures its transition from infectious to replicative forms, revealing a critical dependency on host-derived nutrients.

Together, these studies illustrate that metabolic reprogramming is a bidirectional and context-dependent process that underpins the biology of infection. This chapter provides a comprehensive overview of how pathogens and hosts reshape their metabolic landscapes to gain advantage, offering insights into novel therapeutic strategies targeting metabolic pathways.

6.2 Machine Generated Summaries

Disclaimer: The summaries in this chapter were generated from Springer Nature publications using abstractive AI auto-summarization: An abstraction-based summarizer creates new text based on deep learning. New phrases are created to summarize the content. As the constituted sentences are machine selected, they may not fully reflect the body of the work, so we strongly advise that the original content is read and cited. The auto-generated summaries were curated by the editor to meet Springer Nature publication standards.

Salmonella Typhimurium *Reprograms Macrophage Metabolism via T3SS Effector SopE2 to Promote Intracellular Replication and Virulence [73]*

This is a machine-generated summary of:

Jiang, Lingyan; Wang, Peisheng; Song, Xiaorui; Zhang, Huan; Ma, Shuangshuang; Wang, Jingting; Li, Wanwu; Lv, Runxia; Liu, Xiaoqian; Ma, Shuai; Yan, Jiaqi; Zhou, Haiyan; Huang, Di; Cheng, Zhihui; Yang, Chen; Feng, Lu; Wang, Lei *Salmonella Typhimurium* Reprograms Macrophage Metabolism via T3SS Effector SopE2 to Promote Intracellular Replication and Virulence [73]

Published in: Nature Communications (2021)

Link to original: https://doi.org/10.1038/s41467-021-21186-4

If you want to cite the papers, please refer to the original.

For technical reasons we could not place the page where the original quote is coming from.

Research Highlights

Salmonella enterica serovar Typhimurium (STM) is a significant intracellular pathogen responsible for gastroenteritis and systemic diseases in humans and animals. It invades various cell types, including macrophages, which are crucial for its pathogenesis. Macrophages, typically the first line of defense against pathogens, become a niche for STM replication, enabling systemic infection. Although STM's reliance on macrophage-derived nutrients for replication is known, the impact of macrophage metabolic changes on STM's growth and pathogenicity remains unclear. Now, a new study by Lingyan Jiang and colleagues reveals that STM reprograms macrophage metabolism to enhance its intracellular replication and virulence.

The study employed a combination of metabolomics, transcriptomics, and genetic approaches to investigate how STM alters macrophage metabolism. Researchers infected mouse peritoneal macrophages with STM and observed increased glycolysis and decreased oxidative phosphorylation (OXPHOS), indicative of a Warburg effect. This metabolic shift was confirmed by elevated extracellular acidification rates and reduced oxygen consumption rates. Metabolomic analysis showed significant accumulation of glycolytic intermediates like phosphoglycerate, pyruvate, and lactate in infected macrophages. Transcriptome analysis revealed upregulation of glycolytic enzymes and downregulation of serine synthesis pathway genes in infected cells. The study further demonstrated that STM utilizes these accumulated intermediates as carbon sources to support its replication.

The findings highlight a sophisticated mechanism by which STM exploits host cell metabolism to facilitate its growth. The study identifies the SPI-2 effector as a key player in repressing macrophage serine synthesis, leading to the accumulation of glycolytic intermediates that STM uses for replication. Additionally, the CreBC two-component system senses pyruvate and lactate levels to upregulate SPI-2 gene expression, enhancing STM's virulence. Future studies could explore the potential therapeutic targeting of these metabolic pathways to control STM infections. This research advances our understanding of host–pathogen interactions and opens avenues for developing strategies to combat intracellular bacterial infections.

Reprogramming of Gene Expression in Escherichia coli *Cultured on Pyruvate Versus Glucose [74]*

This is a machine-generated summary of:

Kaberdina, Anna Chao; Ruiz-Larrabeiti, Olatz; Lin-Chao, Sue; Kaberdin, Vladimir R. Reprogramming of Gene Expression in *Escherichia coli* Cultured on Pyruvate Versus Glucose [74]

Published in: Molecular Genetics and Genomics (2019)

Link to original: https://doi.org/10.1007/s00438-019-01597-1

If you want to cite the papers, please refer to the original.

For technical reasons we could not place the page where the original quote is coming from.

Research Highlights

Bacteria have evolved remarkable abilities to adapt and survive in hostile environments, largely due to their capacity to efficiently utilize available resources and adjust their metabolism in response to environmental changes. Among the cellular factors that control bacterial fitness and adaptation, trans-encoded small RNAs (sRNAs) play a crucial role. Previous studies have shown that sRNAs contribute to the regulation of biofilm formation, virulence, and responses to environmental cues such as iron limitation and oxidative stress. Despite extensive research on bacterial carbon metabolism, the regulatory networks controlling gene expression in central carbon metabolism (CCM) remain incompletely defined. Now, a new study by Anna Chao Kaberdina and colleagues explores these networks by comparing gene expression profiles of *Escherichia coli* grown on pyruvate versus glucose.

The study employed a custom microarray developed for simultaneous profiling of protein-coding genes and sRNAs in *E. coli*. Researchers grew *E. coli* K-12 strains in minimal media supplemented with either pyruvate or glucose, allowing them to compare transcriptome profiles under different carbon source conditions. Total RNA was isolated using a modified hot phenol extraction method, followed by microarray analysis to identify differentially expressed genes. The results were further validated using Northern blotting and two-dimensional protein gel electrophoresis. The study revealed that several sRNAs with known or predicted roles in metabolism and stress responses were differentially expressed depending on the carbon source used for growth.

The findings highlight the complexity of gene expression networks involved in *E. coli*'s adaptation to different carbon sources. The study identified major pathways of pyruvate metabolism and suggested that pyruvate conversion to phosphoenolpyruvate might be a principal route for assimilation when pyruvate is the sole carbon source. Additionally, the research uncovered roles for sRNAs such as RyeB, CyaR, GcvB, and RyhB in regulating metabolism, transport, and stress adaptation during growth on alternative carbon sources. Future studies could further elucidate the direct and indirect roles of these sRNAs in controlling CCM at the posttranscriptional level, providing insights into bacterial metabolic engineering applications.

Gut Microbiome and Colon Cancer: Role of Bacterial Metabolites and Their Molecular Targets in the Host [75]

This is a machine-generated summary of:

Bhutia, Yangzom D.; Ogura, Jiro; Sivaprakasam, Sathish; Ganapathy, Vadivel Gut Microbiome and Colon Cancer: Role of Bacterial Metabolites and Their Molecular Targets in the Host [75]

Published in: Current Gastroenterology Reports (2017)

Link to original: https://doi.org/10.1007/s11888-017-0362-9

If you want to cite the papers, please refer to the original.

For technical reasons we could not place the page where the original quote is coming from.

Research Highlights

From the first day of life, humans coexist with a diverse array of microorganisms that colonize various parts of the body. This colonization begins at birth and varies depending on the mode of delivery, with different microbial communities associated with vaginal and Cesarean section births. Traditionally, it was believed that bacterial cells outnumber human cells by a significant margin, but recent studies suggest a closer ratio of 1:1, with most bacteria residing in the large intestine. While these colonic bacteria are often considered commensal, they actually engage in a symbiotic relationship with the host, providing numerous health benefits. However, they can also contribute to colonic and systemic diseases. Now, a new study by Yangzom D. Bhutia and colleagues explores the role of bacterial metabolites as key messengers in the communication between colonic bacteria and their human hosts.

The study delves into how colonic bacteria release specific metabolites into the lumen through various metabolic pathways. These metabolites are influenced by factors such as diet composition and bacterial strains present in the colon. While many metabolites exert local effects on nearby host cells like colonic epithelial cells and mucosal immune cells, some enter systemic circulation, impacting distant organs. The research identifies several groups of bacterial metabolites based on their structure or origin, including short-chain fatty acids like acetate, propionate, and butyrate; tryptophan metabolites; lipid metabolites; and bacterial cell wall components. These metabolites interact with host molecular targets such as G-protein-coupled receptors and nuclear receptors, influencing processes like inflammation, tumor suppression, and immune modulation.

The study advances our understanding of gut microbiota's role in health and disease by highlighting how bacterial metabolites mediate communication between bacteria and host cells. This chemical dialogue is crucial for maintaining colonic health and preventing diseases such as cancer and inflammatory disorders. Future study directions could focus on identifying additional molecular targets for these metabolites and exploring their therapeutic potential in treating systemic diseases linked to gut microbiota. Understanding these interactions may pave the way for novel interventions aimed at modulating gut microbiota to enhance human health.

Metabolic Investigation of Host/Pathogen Interaction Using MS2-Ifected Escherichia coli *[76]*

This is a machine-generated summary of:

Jain, Rishi; Srivastava, Ranjan Metabolic Investigation of Host/Pathogen Interaction Using MS2-Infected *Escherichia coli* [76]

Published in: BMC Systems Biology (2009)

Link to original: https://doi.org/10.1186/1752-0509-3-121

If you want to cite the papers, please refer to the original.

For technical reasons we could not place the page where the original quote is coming from.

Research Highlights

Viruses are responsible for a wide range of diseases, from the common cold to deadly infections like Ebola. They rely on host cells to replicate, causing significant changes in the host's metabolism to favor viral protein production. Understanding these metabolic shifts can reveal potential drug targets. Now, a new study by Rishi Jain and colleagues employs a systems biology approach to explore these changes using a constraints-based modeling technique. This study focuses on the interaction between the bacteriophage Qβ and its host, *Escherichia coli* C-, providing insights into how viral infections alter host metabolism.

The researchers utilized flux balance analysis, a constraints-based modeling approach, to investigate the metabolic changes during the middle steady state of Qβ infection in *E. coli*. They constructed genome-scale metabolic models for both uninfected and infected *E. coli* cells, incorporating experimental data on glucose and oxygen uptake rates. The model for infected cells was modified to include viral proteins and RNA, reflecting the virus's hijacking of host resources for its replication. Linear programming was used to optimize reaction fluxes, with an emphasis on maximizing viral protein production over host growth. The study revealed that viral infection redirects host metabolism toward amino acid biosynthesis and upregulates the pentose phosphate pathway while downregulating the TCA cycle and cell wall synthesis.

This research demonstrates how viral infections can significantly alter host metabolic pathways, providing a framework for identifying potential drug targets. The study highlights similarities between viral infection-induced metabolic changes and those observed in cancer cells, suggesting that this model system could be used to explore treatments targeting altered metabolism in cancer. Future studies could extend these methods to more complex organisms, potentially offering new insights into human cellular metabolism and novel therapeutic strategies against diseases involving metabolic dysregulation.

Host–Microbe Co-metabolism via MCAD Generates Circulating Metabolites Including Hippuric Acid [77]

This is a machine-generated summary of:

Pruss, Kali M.; Chen, Haoqing; Liu, Yuanyuan; Van Treuren, William; Higginbottom, Steven K.; Jarman, John B.; Fischer, Curt R.; Mak, Justin; Wong, Beverly; Cowan, Tina M.; Fischbach, Michael A.; Sonnenburg, Justin L.; Dodd, Dylan Host-Microbe Co-Metabolism via MCAD Generates Circulating Metabolites Including Hippuric Acid [77]

Published in: Nature Communications (2023)

Link to original: https://doi.org/10.1038/s41467-023-36138-3

If you want to cite the papers, please refer to the original.

For technical reasons we could not place the page where the original quote is coming from.

Research Highlights

The human gut microbiome is known to produce a variety of small molecules that can significantly impact human biology. These molecules, derived from the metabolism of dietary and host compounds, can influence physiological processes both locally in the gut and systemically after being absorbed into circulation. Despite growing evidence of the impact of microbial metabolites on host physiology, there remains a significant gap in understanding how these metabolites intersect with host metabolism and their ultimate fate within the body. Now, a new study by Kali M. Pruss and colleagues explores this intersection by investigating how gut microbiota-generated metabolites are processed by host metabolic pathways.

The study utilized gnotobiotic mice colonized with either wild-type *Clostridium sporogenes* or a mutant strain lacking the phenyllactate dehydratase (fldC) gene to explore the role of this gene cluster in metabolite production. Through metabolomic profiling using gas chromatography time-of-flight mass spectrometry, the researchers identified fldC-dependent metabolites in host circulation. They discovered that products of the fld locus undergo β-oxidation via medium chain acyl-CoA dehydrogenase (MCAD) in the host, leading to the production of hippuric acid, one of the most concentrated metabolites found in human urine. The study further demonstrated that phenylpropionic acid, produced by microbial metabolism of phenylalanine, is converted to hippuric acid by the host through this newly identified co-metabolic pathway.

This research uncovers a previously unappreciated mechanism for host–microbe co-metabolism involving β-oxidation and highlights MCAD as a key enzyme in processing microbial metabolites. The findings suggest that variations in gut microbiota can influence circulating metabolite levels, pointing to potential implications for understanding metabolic diseases and developing therapeutic strategies. Future studies could explore other microbial pathways contributing to metabolite production and assess their impact on health outcomes. Additionally, stable isotope tracing could help delineate dietary versus microbial contributions to metabolite levels, offering insights into personalized nutrition and microbiome-targeted therapies.

Metabolomics Deciphered Metabolic Reprogramming Required for Biofilm Formation [78]

This is a machine-generated summary of:

Lu, Haitao; Que, Yumei; Wu, Xia; Guan, Tianbing; Guo, Hao Metabolomics Deciphered Metabolic Reprogramming Required for Biofilm Formation [78]

Published in: Scientific Reports (2019)

Link to original: https://doi.org/10.1038/s41598-019-49603-1

If you want to cite the papers, please refer to the original.

For technical reasons we could not place the page where the original quote is coming from.

Research Highlights

Bacterial biofilms are complex structures composed of a self-produced matrix of extracellular polymeric substances, including extracellular DNA, proteins, and polysaccharides. These biofilms enable bacterial cells to survive and persist, playing a crucial role in many human infections due to their high resistance to antibiotics. Urinary tract infections (UTIs), often caused by uropathogenic *Escherichia coli* (UPEC), are common and recurrent, with UPEC forming biofilms that make them difficult to eradicate with antibiotics. Despite understanding the stages of biofilm formation, the role of small-molecule metabolites in this process remains unclear. Now, a new study by Haitao Lu and colleagues investigates the metabolic reprogramming associated with UPEC biofilm formation using a combination of staining assays, cytological observations, and metabolomics methods.

The study utilized a clinical UPEC strain capable of forming a floating pellicle-biofilm in YESCA medium. Researchers employed crystal violet staining to quantify biofilm formation and used scanning electron microscopy (SEM) and transmission electron microscopy (TEM) to observe the intricate structure of the biofilm. Metabolomics analysis was conducted using gas chromatography-mass spectrometry (GC-MS) and liquid chromatography-mass spectrometry (LC-MS) to profile small-molecule metabolites. The results revealed significant metabolic reprogramming during biofilm formation compared to planktonic populations. Differential metabolites identified included amino acids, carbohydrates, organic acids, glycerol-derived compounds, polyamines, and uridine. These findings suggest that metabolic changes are critical for biofilm formation and may influence the expression of virulence factors.

The study advances our understanding of biofilm formation by highlighting the importance of metabolic reprogramming in this process. The identification of key metabolites and pathways involved in UPEC biofilm formation provides potential targets for developing new treatments against UTIs. Future studies could explore the mechanisms by which these metabolites influence biofilm development and investigate their roles in other bacterial strains. This research offers valuable insights into the metabolic underpinnings of biofilm formation, paving the way for novel therapeutic strategies to combat persistent bacterial infections.

Staphylococcus aureus *Host Interactions and Adaptation [79]*

This is a machine-generated summary of:

Howden, Benjamin P.; Giulieri, Stefano G.; Wong Fok Lung, Tania; Baines, Sarah L.; Sharkey, Liam K.; Lee, Jean Y. H.; Hachani, Abderrahman; Monk, Ian R.; Stinear, Timothy P. *Staphylococcus aureus* Host Interactions and Adaptation [79]

Published in: Nature Reviews Microbiology (2023)

Link to original: https://doi.org/10.1038/s41579-023-00852-y

If you want to cite the papers, please refer to the original.

For technical reasons we could not place the page where the original quote is coming from.

Research Highlights

Staphylococcus aureus is a prevalent colonizer of humans and a leading opportunistic pathogen, responsible for significant morbidity and mortality worldwide. Despite its longstanding interaction with humans, *S. aureus* remains a formidable challenge due to its ability to evade the immune system and develop resistance to antibiotics. The complexity of its interactions with the host microbiota and immune system underscores the need for a deeper understanding of its pathophysiology. Now, a new study by Benjamin P. Howden and colleagues explores recent advances in understanding the mechanisms *S. aureus* employs to cause infections, focusing on immune evasion, biofilm formation, and genomic plasticity.

The study utilizes genomics-enabled microbial community analyses and metabolomics to investigate the interactions between *S. aureus* and other commensals during colonization. The researchers highlight the role of biofilms in chronic infections, noting that biofilm-associated infections are often linked to increased antimicrobial resistance and immune evasion. By examining genetic diversity within the *S. aureus* genome, the study reveals how point mutations and recombination contribute to adaptation within hosts, particularly during infection transitions. The authors also explore how *S. aureus* adapts through metabolic changes that influence host–pathogen interactions, such as shifting macrophage metabolism toward an anti-inflammatory state to promote bacterial persistence.

The study's findings emphasize the importance of understanding *S. aureus*'s adaptive mechanisms in developing effective therapeutic strategies. Future study directions include exploring immunometabolic therapies that target host–pathogen metabolic interactions and leveraging pathogen genomics for patient management. The research suggests that integrating insights from human genomics and multi-omics profiling could identify targeted therapeutics for severe staphylococcal diseases. By advancing our understanding of *S. aureus*'s genetic variability and adaptation strategies, this research paves the way for novel prevention strategies and therapeutic approaches that address the challenges posed by this versatile pathogen.

Microbial Metabolites Regulate Host Lipid Metabolism Through NR5A–Hedgehog Signalling [80]

This is a machine-generated summary of:

Lin, Chih-Chun Janet; Wang, Meng C. Microbial Metabolites Regulate Host Lipid Metabolism Through NR5A–Hedgehog Signalling [80]

Published in: Nature Cell Biology (2017)

Link to original: https://doi.org/10.1038/ncb3515

If you want to cite the papers, please refer to the original.

For technical reasons we could not place the page where the original quote is coming from.

Research Highlights

Microorganisms are known to live in close association with eukaryotic organisms, influencing physiological homeostasis and disease susceptibility across the plant and animal kingdoms. These microorganisms can rapidly alter their transcriptomes, proteomes, and biochemical profiles in response to environmental changes, affecting host metabolism. However, the molecular mechanisms linking environmental inputs, microorganism-derived metabolites, and host signaling pathways in regulating host lipid metabolism remain poorly understood. Now, a new study by Chih-Chun Janet Lin and colleagues investigates this relationship using the *Escherichia coli–Caenorhabditis elegans* system.

The researchers cultured *C. elegans* on *E. coli* grown in either LB medium or minimal salts medium to explore how different bacterial environments affect host lipid metabolism. They found that *C. elegans* grown on bacteria from minimal salts medium had significantly higher fat content than those grown on LB medium. This increase in fat content was independent of developmental exposure and occurred rapidly within hours of switching bacterial environments. The study utilized label-free lipid imaging techniques such as stimulated Raman scattering microscopy to quantify these changes and confirmed them biochemically. Further analysis revealed that the increased fat storage was not due to global nutritional differences between the bacteria but was actively regulated by specific bacterial metabolites, particularly those related to methionine deprivation.

The study advances our understanding of how bacterial metabolic activities influence host lipid metabolism by identifying methionine deprivation as a key environmental input that triggers bacterial metabolic alterations affecting host metabolism. Future studies could explore the roles of mitochondria in bacteria–host interactions for improving host fitness, given the evolutionary relationship between mitochondria and bacteria. The findings highlight the complexity of metabolic adaptation responses at the organismal level and suggest potential avenues for further research into microbial influences on host physiology.

Humoral Regulation of Iron Metabolism by Extracellular Vesicles Drives Antibacterial Response [81]

This is a machine-generated summary of:

Kuang, Huijuan; Dou, Geng; Cheng, Linfeng; Wang, Xiangdong; Xu, Haokun; Liu, Xuemei; Ding, Feng; Yang, Xiaoshan; Liu, Siying; Bao, Lili; Liu, Huan; Liu, Yao; Li, Bei; Jin, Yan; Liu, Shiyu Humoral Regulation of Iron Metabolism by Extracellular Vesicles Drives Antibacterial Response [81]

Published in: Nature Metabolism (2023)

Link to original: https://doi.org/10.1038/s42255-022-00723-5

If you want to cite the papers, please refer to the original.

For technical reasons we could not place the page where the original quote is coming from.

Research Highlights

Bacterial infections can lead to severe conditions such as bacteremia and sepsis, posing significant threats to human health. Iron plays a crucial role in various biological processes, and the host's immediate response to bacterial invasion is to restrict iron availability, a defense mechanism known as nutritional immunity. Previous studies have shown that liver- and spleen-resident macrophages sequester iron by capturing iron-containing proteins. However, these studies primarily focus on cell-dependent regulation of iron metabolism. Now, a new study by Huijuan Kuang and colleagues reveals a previously unknown humoral regulatory mechanism involving extracellular vesicles (EVs) that act as "iron catchers" during bacterial infection, thereby broadening our understanding of iron metabolism.

The study utilized a mouse model infected with *Salmonella Typhimurium* to investigate the role of EVs in iron sequestration during bacterial infection. Researchers observed an increase in EV concentration in the serum following infection, suggesting that EV release is part of the host's response. By labeling EVs and tracking their distribution, it was found that they predominantly homed to the liver and spleen, aligning with systemic iron distribution changes during infection. The team further explored the impact of blocking EV release using pharmacological agents, which resulted in disrupted iron recycling and worsened infection outcomes. Additionally, they demonstrated that EVs derived from infected macrophages could bind iron-containing proteins through receptors like transferrin receptor (TfR), preventing bacterial access to iron and enhancing host defense.

The findings highlight a novel humoral mechanism where EVs facilitate rapid iron sequestration during bacterial infections, offering insights into nutritional immunity. Future studies could explore the potential therapeutic applications of EVs in combating infections by enhancing their "iron-catching" capabilities or developing strategies to modulate their release. The study also suggests investigating other types of extracellular vesicles and their roles in nutrient regulation during infections. These advances underscore the importance of understanding both cellular and humoral aspects of iron metabolism in developing effective treatments against bacterial pathogens.

Reprogramming of Host Glutamine Metabolism During **Chlamydia trachomatis** *Infection and Its Key Role in Peptidoglycan Synthesis [82]*

This is a machine-generated summary of:

Rajeeve, Karthika; Vollmuth, Nadine; Janaki-Raman, Sudha; Wulff, Thomas F.; Baluapuri, Apoorva; Dejure, Francesca R.; Huber, Claudia; Fink, Julian; Schmalhofer, Maximilian; Schmitz, Werner; Sivadasan, Rajeeve; Eilers, Martin; Wolf, Elmar; Eisenreich, Wolfgang; Schulze, Almut; Seibel, Jürgen; Rudel, Thomas Reprogramming of Host Glutamine Metabolism During *Chlamydia trachomatis* Infection and Its Key Role in Peptidoglycan Synthesis [82]

Published in: Nature Microbiology (2020)

Link to original: https://doi.org/10.1038/s41564-020-0762-5

If you want to cite the papers, please refer to the original.

For technical reasons we could not place the page where the original quote is coming from.

Research Highlights

Chlamydia trachomatis, an obligate intracellular bacterium, is the leading cause of sexually transmitted bacterial infections. It undergoes a biphasic developmental cycle, transitioning between elementary bodies (EBs) and reticulate bodies (RBs) within host cells. While *Chlamydia* relies heavily on host-derived metabolites due to its reduced genome, the mechanisms triggering the transition from EBs to RBs remain unclear. Now, a new study by Karthika Rajeeve and colleagues reveals that glutamine (Gln) plays a crucial role in this transition by serving as a carbon source for peptidoglycan biosynthesis and initiating the EB-to-RB conversion.

The study employed metabolite-uptake assays in axenic culture to investigate how *Chlamydia* utilizes Gln. Using isotopologue profiling with labeled Gln, the researchers demonstrated that Gln is metabolized via glutaminolysis and the truncated TCA cycle to produce intermediates like glutamate and aspartate, essential for peptidoglycan synthesis. Electron microscopy showed that EBs incubated with Gln underwent morphological changes toward RBs, termed intermediate bodies, and exhibited reduced infectivity. Additionally, RNA sequencing revealed that Gln exposure activated early genes associated with the EB-to-RB transition. The study also explored how *Chlamydia* infection alters host cell metabolism, finding increased Gln uptake and metabolism in infected cells, driven by upregulated c-Myc levels.

This research advances our understanding of *Chlamydia*'s metabolic interactions with host cells and highlights Gln's pivotal role in its developmental cycle. Future studies could explore therapeutic strategies targeting Gln metabolism to combat *Chlamydia* infections. The findings suggest that disrupting Gln uptake or metabolism might hinder *Chlamydia*'s ability to replicate and develop within host cells. This approach could potentially lead to novel treatments for infections caused by this pathogen, leveraging insights into its metabolic dependencies on host-derived nutrients.

Staphylococcus aureus *Small Colony Variants Impair Host Immunity by Activating Host Cell Glycolysis and Inducing Necroptosis [83]*

This is a machine-generated summary of:

Wong Fok Lung, Tania; Monk, Ian R.; Acker, Karen P.; Mu, Andre; Wang, Nancy; Riquelme, Sebastián A.; Pires, Silvia; Noguera, Loreani P.; Dach, Felix; Gabryszewski, Stanislaw J.; Howden, Benjamin P.; Prince, Alice *Staphylococcus aureus* Small Colony Variants Impair Host Immunity by Activating Host Cell Glycolysis and Inducing Necroptosis [83]

Published in: Nature Microbiology (2019)

Link to original: https://doi.org/10.1038/s41564-019-0597-0

If you want to cite the papers, please refer to the original.

For technical reasons we could not place the page where the original quote is coming from.

Research Highlights

Staphylococcus aureus is a versatile human pathogen known for causing a range of infections, from acute sepsis to chronic conditions like endocarditis and osteomyelitis. While the role of toxins in the early stages of infection is well-documented, the mechanisms by which *S. aureus* persists in chronic infections remain less understood. Small colony variants (SCVs) of *S. aureus*, characterized by mutations affecting their metabolic pathways, are often implicated in these persistent infections. However, the exact process by which SCVs are selected and maintained during chronic infection is unclear. Now, a new study by Tania Wong Fok Lung and colleagues explores how SCVs influence host cell metabolism to promote their persistence.

The study utilized a ΔhemB mutant SCV in the clinically relevant LAC background to investigate its interaction with host cells. The researchers constructed and characterized this mutant alongside its complemented strain to ensure accurate comparisons. They found that both wild-type *S. aureus* and the ΔhemB SCV induced glycolysis and mitochondrial reactive oxygen species (ROS) production in host keratinocytes, leading to necroptosis—a form of programmed cell death that releases viable bacteria and aids SCV persistence. The ΔhemB SCV also upregulated bacterial fumC expression, reducing local fumarate levels and inhibiting trained immunity, which is crucial for protection against secondary infections. These findings were supported by metabolomic analyses showing distinct metabolic responses in host cells infected with SCVs compared to wild-type strains.

The study advances our understanding of *S. aureus* pathogenesis by highlighting the role of metabolic interactions between SCVs and host cells in promoting chronic infection. The ability of SCVs to induce necroptosis without relying on toxin production suggests alternative pathways for bacterial persistence. Future studies could focus on targeting these metabolic pathways to develop new therapeutic strategies against chronic *S. aureus* infections. Additionally, understanding how SCVs evade trained immunity could inform vaccine development efforts aimed at preventing recurrent staphylococcal infections.

6.3 Conclusion

The studies presented in this chapter show the profound impact of metabolic reprogramming on the dynamics of bacterial infections. Far from being passive biochemical processes, metabolic pathways are actively reshaped by both pathogens and host cells in response to infection. These reprogramming events serve as strategic adaptations that influence pathogen survival, host defense, immune modulation, and disease progression. By examining the molecular mechanisms underlying these shifts, this chapter reveals metabolism as a central axis in the host–pathogen interface—one that offers promising targets for therapeutic intervention.

One of the most striking themes is the ability of bacterial pathogens to manipulate host metabolism to create favorable niches for replication. *S. Typhimurium*, for example, reprograms macrophage metabolism by inducing a Warburg-like effect, increasing glycolysis and suppressing oxidative phosphorylation. This shift not only provides the pathogen with essential carbon sources but also dampens host immune responses, facilitating intracellular survival. The pathogen's use of the CreBC two-component system to sense host-derived metabolites and regulate virulence gene expression exemplifies the precision with which bacteria exploit host metabolic states.

Similarly, *S. aureus* small colony variants (SCVs) rewire host cell metabolism to promote persistence. By inducing glycolysis and necroptosis in keratinocytes, SCVs evade immune clearance and establish chronic infections. These findings highlight alternative virulence strategies that bypass traditional toxin-mediated mechanisms, suggesting that metabolic manipulation is a key driver of bacterial persistence in chronic disease contexts.

The role of microbial metabolites in shaping host physiology is another critical insight. In the gut, bacterial metabolites such as short-chain fatty acids, tryptophan derivatives, and lipid components interact with host receptors to modulate inflammation, immunity, and even tumor suppression. These metabolites act as chemical messengers, linking microbial activity to host cellular responses. In the context of colon cancer, this metabolic dialogue influences disease risk and progression, underscoring the therapeutic potential of targeting microbial metabolism.

Host–microbe co-metabolism further illustrates the integration of microbial and host metabolic pathways. The discovery that microbial metabolites are processed by host enzymes like MCAD to produce circulating compounds such as hippuric acid reveals a complex interplay that extends beyond the gut. These co-metabolic products reflect the combined activities of host and microbiota, offering biomarkers for health and disease and informing personalized nutrition strategies.

Viral infections also induce metabolic reprogramming, as demonstrated in bacteriophage-infected *E. coli.* Systems biology approaches reveal that viral replication redirects host metabolic fluxes toward amino acid biosynthesis and the pentose phosphate pathway, while suppressing the TCA cycle. These shifts mirror those seen in cancer cells, suggesting shared strategies among diverse pathogens to hijack host resources. Such insights could inform the development of antiviral therapies targeting metabolic vulnerabilities.

Biofilm formation, a hallmark of chronic bacterial infections, is shown to be driven by metabolic reprogramming. In uropathogenic *E. coli*, distinct shifts in amino acids, carbohydrates, and polyamines support biofilm development and antibiotic resistance. Understanding these metabolic changes provides a foundation for designing interventions that disrupt biofilm integrity and enhance treatment efficacy.

Host responses to infection also involve sophisticated metabolic adaptations. The discovery that extracellular vesicles (EVs) act as iron catchers during *Salmonella* infection reveals a novel humoral mechanism of nutritional immunity. By sequestering iron, EVs limit bacterial access to this essential nutrient, enhancing host defense. This finding expands our understanding of systemic immune responses and suggests new avenues for therapeutic modulation of iron metabolism.

Finally, *Chlamydia trachomatis* exemplifies how intracellular pathogens depend on host metabolic reprogramming for their developmental cycles. By hijacking host glutamine metabolism, *Chlamydia* ensures its transition from infectious to replicative forms and supports peptidoglycan synthesis. Targeting these metabolic dependencies could offer novel strategies for controlling infections caused by obligate intracellular bacteria.

In conclusion, metabolic reprogramming is a dynamic and bidirectional process that shapes the outcome of bacterial infections. The studies in this chapter reveal that both pathogens and hosts actively remodel their metabolism to gain advantage, influencing virulence, immune responses, and disease progression. As our understanding of these processes deepens, new opportunities will emerge for developing targeted therapies that disrupt pathogenic metabolic strategies or bolster host defenses. Metabolism, once considered a background player, now stands at the forefront of infection biology and therapeutic innovation.

Bibliography

1. Basan M, Hui S, Okano H, Zhang Z, Shen Y, Williamson JR, Hwa T (2015) Overflow metabolism in *Escherichia coli* results from efficient proteome allocation. Nature 528(7580):99–104. https://doi.org/10.1038/nature15765
2. Cao Y, Forstner KU, Vogel J, Smith CJ (2016) Cis-encoded small RNAs, a conserved mechanism for repression of polysaccharide utilization in bacteroides. J Bacteriol 198(18):2410–2418. https://doi.org/10.1128/JB.00381-16
3. Morin M, Ropers D, Letisse F, Laguerre S, Portais JC, Cocaign-Bousquet M, Enjalbert B (2016) The post-transcriptional regulatory system CSR controls the balance of metabolic pools in upper glycolysis of *Escherichia coli*. Mol Microbiol 100(4):686–700. https://doi.org/10.1111/mmi.13343
4. Richards GR, Vanderpool CK (2011) Molecular call and response: the physiology of bacterial small RNAs. Biochim Biophys Acta 1809(10):525–531. https://doi.org/10.1016/j.bbagrm.2011.07.013
5. Michaux C, Verneuil N, Hartke A, Giard JC (2014) Physiological roles of small RNA molecules. Microbiology 160(Pt 6):1007–1019. https://doi.org/10.1099/mic.0.076208-0
6. Ruiz-Larrabeiti O, Plagaro AH, Gracia C, Sevillano E, Gallego L, Hajnsdorf E, Kaberdin VR (2016) A new custom microarray for sRNA profiling in *Escherichia coli*. FEMS Microbiol Lett. https://doi.org/10.1093/femsle/fnw131
7. Sender, R., Fuchs, S. & Milo, R. Revised estimates for the number of human and bacteria cells in the body. *PLoS Biol. 14*, e1002533 1–21 (2016).
8. • Schroeder BO, Backhed F. Signals from the gut microbiota to distant organs in physiology and disease. Nat Med. 2016;22:1079–89. *This is an outstanding review of the literature summarizing the signals from the gut bacteria that travel to distant organs and influence their function in health and disease.*
9. Bhutia YD, Ganapathy V. Short, but smart: SCFAs train T cells in the gut to fight autoimmunity in the brain. Immunity. 2015;43:629–31.
10. •• Haghikia A, Jorg S, Duscha A, et al. Dietary fatty acids directly impact central nervous system autoimmunity via the small intestine. Immunity. 2015;43:817–29. *This study demonstrates that dietary fatty acids, by modulating gut microbiota and their metabolism, regulate immune cells in the gut, which then travel to the periphery to impact systemic immunity.*

11. Tazoe H, Otomo Y, Kaji I, et al. Roles of short-chain fatty acids receptors, GPR41 and GPR43 on colonic functions. J Physiol Pharmacol. 2008;59 Suppl 2:252–62.
12. Ganapathy V, Thangaraju M, Prasad PD, Martin PM, Singh N. Transporters and receptors for short-chain fatty acids as the molecular link between colonic bacteria and the host. Curr Opin Pharmacol. 2013;13:869–74.
13. Tan J, McKenzie C, Potamitis M, Thorburn AN, Mackay CR, Macia L. The role of short-chain fatty acids in health and disease. Adv Immunol. 2014;121:91–119.
14. Cresci GA, Thangaraju M, Mellinger JD, Liu K, Ganapathy V. Colonic gene expression in conventional and germ-free ice with a focus on the butyrate receptor GPR109A and the butyrate transporter SLC5A8. J Gastrointest Sug. 2010;14:449–61.
15. •• Singh N, Gurav A, Sivaprakasam S, et al. Activation of Gpr109a, receptor for niacin and the commensal metabolite butyrate, suppresses colonic inflammation and carcinogenesis. Immunity. 2014;40:128–39. *This study demonstrated unequivocally the anti-inflammatory and tumor-suppressive function of the butyrate receptor GPR109A in vivo using Gpr109a-null mice.*
16. Liu C, Wu J, Zhu J, et al. Lactate inhibits lipolysis in fat cells through activation of an orphan G-protein-coupled receptor. J Biol Chem. 2009;284:2811–22.
17. He W, Miao FJ, Lin DC, et al. Citric acid cycle intermediates as ligands for orphan G-protein-coupled receptors. Nature. 2004;429:188–93.
18. Hubbard, T. D., Murray, I. A. & Perdew, G. H. Indole and tryptophan metabolism: endogenous and dietary routes to ah receptor activation. *Drug Metab. Dispos. 43*, 1522–1535 (2015).
19. Jin UH, Lee SO, Sridharan G, et al. Microbiome-derived tryptophan metabolites and their aryl hydrocarbon receptor-dependent agonist and antagonist activities. Mol Pharmacol. 2014;85:777–88.
20. Wikoff, W. R. et al. Metabolomics analysis reveals large effects of gut microflora on mammalian blood metabolites. *Proc. Natl Acad. Sci. USA 106*, 3698–3703 (2009).
21. Venkatesh M, Mukherjee S, Wang H, et al. Symbiotic bacterial metabolites regulate gastrointestinal barrier function via the xenobiotic sensor PXR and Toll-like receptor 4. Immunity. 2014;41:296–310.
22. Stockinger B, Di Meglio P, Gialitakis M, Duarte JH. The aryl hydrocarbon receptor: multitasking in the immune system. Annu Rev Immunol. 2014;32:403–32.
23. Xie G, Raufman JP. Role of the aryl hydrocarbon receptor in colon neoplasia. Cancers. 2015;7:1436–46.
24. Goettel JA, Gandhi R, Kenison JE, et al. AHR activation is protective against colitis driven by T cells in humanized mice. Cell Rep. 2016;17:1318–29.
25. Diaz-Diaz CJ, Ronnekleiv-Kelly SM, Nukaya M, Geiger PG, et al. The aryl hydrocarbon receptor is a repressor of inflammation-associated colorectal tumorigenesis in mouse. Ann Surg. 2016;264:429–36.
26. Pondugula SR, Pavek P, Mani S. Pregnane X receptor and cancer: context-specificity is key. Nucl Receptor Res. 2016;3.
27. •• Koeth RA, Wang Z, Levison BS, et al. Intestinal microbiota metabolism of L-carnitine, a nutrient in red meat, promotes atherosclerosis. Nat Med. 2013;19:576–85. *This was the first report on the generation of trimethylamine (TMA) in the colon by bacterial metabolism of diet-derived carnitine and the subsequent hepatic conversion of TMA into the cardiovascular toxin TMA oxide, thus providing a mechanistic link between high dietary intake of carnitine-rich red meat and risk for cardiovascular disease.*
28. • Tang WH, Wang Z, Levison BS, et al. Intestinal microbial metabolism of phosphatidylcholine and cardiovascular risk. N Engl J Med. 2013;368:1575–84. *This study provides evidence for the association between increased plasma levels of TMAO and risk for cardiovascular disease and also for dietary lipid phosphatidylcholine as the source for colonic bacterial metabolism to generate TMA, the precursor for TMAO.*
29. •• Kuo WT, Lee TC, Yang HY, et al. LPS receptor subunits have antagonistic roles in epithelial apoptosis and colonic carcinogenesis. Cell Death Diff. 2015;22:1590–604. *This study provided*

the molecular rationale for the antagonistic actions of LPS in colonic epithelium via the presence or absence of the co-expression of CD14 with TLR4.
30. • Janelsins BM, Lu M, Datta SK. Altered inactivation of commensal LPS due to acyloxyacyl hydrolase deficiency in colonic dendritic cells impairs mucosal Th17 immunity. Proc Natl Acad Sci U S A. 2014;111:373–8. *This study describes the role of chronic exposure of the colon to LPS in the polarization of naïve T cells towards pro-inflammatory Th17-positive T cells or immunosuppressive regulatory T cells (Tregs).*
31. Ikeda M, Kato N: Modulation of host metabolism as a target of new antivirals. Advanced Drug Delivery Reviews. 2007, 59: 1277–1289. https://doi.org/10.1016/j.addr.2007.03.021
32. van Duin J, Tsareva N: The Single-Stranded RNA Bacteriophages. 1988, New York: Plenum Press, 1
33. Berzin V, Rosenthal G, Gren EJ: Cellular macromolecule synthesis in Escherichia coli infected with bacteriophage MS2. Eur J Biochem. 1974, 45: 233–242. https://doi.org/10.1111/j.1432-1033.1974.tb03547.x
34. Feist AM, Henry CS, Reed JL, Krummenacker M, Joyce AR, Karp PD, Broadbelt LJ, Hatzimanikatis V, Palsson BO: A genome-scale metabolic reconstruction for Escherichia coli K-12 MG1655 that accounts for 1260 ORFs and thermodynamic information. Mol Syst Biol. 2007, 3: 121 https://doi.org/10.1038/msb4100155
35. Mair, R. D., Sirich, T. L., Plummer, N. S. & Meyer, T. W. Characteristics of colon-derived uremic solutes. *Clin. J. Am. Soc. Nephrol. 13*, 1398–1404 (2018).
36. Badenhorst, C. P. S., Erasmus, E., van der Sluis, R., Nortje, C. & van Dijk, A. A. A new perspective on the importance of glycine conjugation in the metabolism of aromatic acids. *Drug Metab. Rev. 46*, 343–361 (2014).
37. Nemet, I. et al. A cardiovascular disease-linked gut microbial metabolite acts via adrenergic receptors. *Cell 180*, 862–877 (2020).
38. Costerton, J. W., Lewandowski, Z., Caldwell, D. E., Korber, D. R. & Lappin-Scott, H. M. Microbial biofilms. *Annual review of microbiology 49*, 711–745, https://doi.org/10.1146/annurev.mi.49.100195.003431 (1995).
39. Hung, C. *et al. Escherichia coli* biofilms have an organized and complex extracellular matrix structure. *mBio 4*, e00645–00613, https://doi.org/10.1128/mBio.00645-13 (2013).
40. Lv, H., Hung, C. S. & Henderson, J. P. Metabolomic analysis of siderophore cheater mutants reveals metabolic costs of expression in uropathogenic *Escherichia coli. Journal of proteome research 13*, 1397–1404, https://doi.org/10.1021/pr4009749 (2014).
41. Yan, L., Nie, W. & Lv, H. Metabolic phenotyping of the Yersinia high-pathogenicity island that regulates central carbon metabolism. *The Analyst 140*, 3356–3361, https://doi.org/10.1039/c4an02223h (2015).
42. Kendall, M. M. & Sperandio, V. Cell-to-Cell Signaling in *Escherichia coli* and Salmonella. *EcoSal Plus 6*, https://doi.org/10.1128/ecosalplus.ESP-0002-2013 (2014).
43. Montgomery, C. P., David, M. Z. & Daum, R. S. Host factors that contribute to recurrent staphylococcal skin infection. *Curr. Opin. Infect. Dis. 28*, 253–258 (2015).
44. Kallen, A. J. et al. Health care-associated invasive MRSA infections, 2005–2008. *JAMA 304*, 641–648 (2010).
45. Giulieri, S. G. et al. Niche-specific genome degradation and convergent evolution shaping *Staphylococcus aureus* adaptation during severe infections. *eLife 11*, e77195 (2022). *Population genomics conducted on a very large clinical cohort to reveal with unprecedented resolution the bacterial genome-wide changes that are associated with the transition from colonizing to invasive S. aureus infections.*
46. Laabei, M. et al. Evolutionary trade-offs underlie the multi-faceted virulence of *Staphylococcus aureus. PLoS Biol. 13*, e1002229 (2015).
47. Stapels, D. A. et al. *Staphylococcus aureus* protects its immune-evasion proteins against degradation by neutrophil serine proteases. *Cell Microbiol. 18*, 536–545 (2016).
48. Askarian, F. et al. Serine-aspartate repeat protein D increases *Staphylococcus aureus* virulence and survival in blood. *Infect. Immun. 85*, e00559–16 (2017).

49. Bhattacharya, M. et al. Leukocidins and the nuclease nuc prevent neutrophil-mediated killing of *Staphylococcus aureus* biofilms. *Infect. Immun. 88*, e00372–20 (2020).
50. Heim, C. E. et al. Myeloid-derived suppressor cells contribute to *Staphylococcus aureus* orthopedic biofilm infection. *J. Immunol. 192*, 3778–3792 (2014).
51. Heim, C. E. et al. Lactate production by *Staphylococcus aureus* biofilm inhibits HDAC11 to reprogramme the host immune response during persistent infection. *Nat. Microbiol. 5*, 1271–1284 (2020). *Sophisticated host–pathogen study using at-scale transposon mutagenesis to reveal the role mechanism of S. aureus lactate in regulation of IL-10 expression.*
52. Tomlinson, K. L. & Riquelme, S. A. Host–bacteria metabolic crosstalk drives *S. aureus* biofilm. *Microb. Cell 8*, 106–107 (2021).
53. Tomlinson, K. L. et al. *Staphylococcus aureus* induces an itaconate-dominated immunometabolic response that drives biofilm formation. *Nat. Commun. 12*, 1399 (2021).
54. Acker, K. P. et al. Strains of *Staphylococcus aureus* that colonize and infect skin harbor mutations in metabolic genes. *iScience 19*, 281–290 (2019).
55. Wong Fok Lung, T. et al. *Staphylococcus aureus* small colony variants impair host immunity by activating host cell glycolysis and inducing necroptosis. *Nat. Microbiol. 5*, 141–153 (2020).
56. Gabryszewski, S. J. et al. Metabolic adaptation in methicillin-resistant *Staphylococcus aureus* pneumonia. *Am. J. Respir. Cell Mol. Biol. 61*, 185–197 (2019).
57. Sheppard, S. K., Guttman, D. S. & Fitzgerald, J. R. Population genomics of bacterial host adaptation. *Nat. Rev. Genet. 19*, 549–565 (2018). *Comprehensive review of bacterial host species adaptation.*
58. Richardson, E. J. et al. Gene exchange drives the ecological success of a multi-host bacterial pathogen. *Nat. Ecol. Evol. 2*, 1468–1478 (2018). *Excellent demonstration of bacterial population genomics to reveal the extent to which humans are shaping the evolution of S. aureus and exchange between different animal hosts.*
59. Bacigalupe, R., Tormo-Mas, M. A., Penades, J. R. & Fitzgerald, J. R. A multihost bacterial pathogen overcomes continuous population bottlenecks to adapt to new host species. *Sci. Adv. 5*, eaax0063 (2019). *Experimental study demonstrating the capacity of S. aureus to acquire beneficial mutations alleviating evolutionary bottlenecks and enabling its adaptation to different hosts.*
60. Drakesmith, H. & Prentice, A. M. Hepcidin and the iron-infection axis. *Science 338*, 768–772 (2012).
61. Cassat, J. E. & Skaar, E. P. Iron in infection and immunity. *Cell Host Microbe 13*, 509–519 (2013).
62. Verweij, F. J. et al. Live tracking of inter-organ communication by endogenous exosomes in vivo. *Dev. Cell 48*, 573–589 (2019).
63. Villarroya-Beltri, C. et al. ISGylation controls exosome secretion by promoting lysosomal degradation of MVB proteins. *Nat. Commun. 7*, 13588 (2016).
64. Jeppesen, D. K. et al. Reassessment of exosome composition. *Cell 177*, 428–445 (2019).
65. Schorey, J. S., Cheng, Y., Singh, P. P. & Smith, V. L. Exosomes and other extracellular vesicles in host–pathogen interactions. *EMBO Rep. 16*, 24–43 (2015).
66. Stephens, R. S. et al. Genome sequence of an obligate intracellular pathogen of humans: *Chlamydia trachomatis*. *Science 282*, 754–759 (1998).
67. Mehlitz, A. et al. Metabolic adaptation of *Chlamydia trachomatis* to mammalian host cells. *Mol. Microbiol. 103*, 1004–1019 (2017).
68. Omsland, A., Sager, J., Nair, V., Sturdevant, D. E. & Hackstadt, T. Developmental stage-specific metabolic and transcriptional activity of *Chlamydia trachomatis* in an axenic medium. *Proc. Natl Acad. Sci. USA 109*, 19781–19785 (2012).
69. Kahl, B. C., Becker, K. & Loffler, B. Clinical significance and pathogenesis of staphylococcal small colony variants in persistent infections. *Clin. Microbiol. Rev. 29*, 401–427 (2016).
70. Proctor, R. A. & Peters, G. Small colony variants in staphylococcal infections: diagnostic and therapeutic implications. *Clin. Infect. Dis. 27*, 419–422 (1998).

71. Kitur, K. et al. Necroptosis promotes *Staphylococcus aureus* clearance by inhibiting excessive inflammatory signaling. *Cell Rep. 16*, 2219–2230 (2016).
72. Arts, R. J., Joosten, L. A. & Netea, M. G. Immunometabolic circuits in trained immunity. *Semin. Immunol. 28*, 425–430 (2016).
73. Jiang, Lingyan; Wang, Peisheng; Song, Xiaorui; Zhang, Huan; Ma, Shuangshuang; Wang, Jingting; Li, Wanwu; Lv, Runxia; Liu, Xiaoqian; Ma, Shuai; Yan, Jiaqi; Zhou, Haiyan; Huang, Di; Cheng, Zhihui; Yang, Chen; Feng, Lu; Wang, Lei. Salmonella Typhimurium reprograms macrophage metabolism via T3SS effector SopE2 to promote intracellular replication and virulence. Nature Communications (2021). https://doi.org/10.1038/s41467-021-21186-4
74. Kaberdina, Anna Chao; Ruiz-Larrabeiti, Olatz; Lin-Chao, Sue; Kaberdin, Vladimir R. Reprogramming of gene expression in Escherichia coli cultured on pyruvate versus glucose. Molecular Genetics and Genomics (2019). https://doi.org/10.1007/s00438-019-01597-1
75. Bhutia, Yangzom D.; Ogura, Jiro; Sivaprakasam, Sathish; Ganapathy, Vadivel Gut Microbiome and Colon Cancer: Role of Bacterial Metabolites and Their Molecular Targets in the Host. Current Gastroenterology Reports (2017). https://doi.org/10.1007/s11888-017-0362-9
76. Jain, Rishi; Srivastava, Ranjan. Metabolic investigation of host/pathogen interaction using MS2-infected Escherichia coli. BMC Systems Biology (2009). https://doi.org/10.1186/1752-0509-3-121
77. Pruss, Kali M.; Chen, Haoqing; Liu, Yuanyuan; Van Treuren, William; Higginbottom, Steven K.; Jarman, John B.; Fischer, Curt R.; Mak, Justin; Wong, Beverly; Cowan, Tina M.; Fischbach, Michael A.; Sonnenburg, Justin L.; Dodd, Dylan. Host-microbe co-metabolism via MCAD generates circulating metabolites including hippuric acid. Nature Communications (2023). https://doi.org/10.1038/s41467-023-36138-3
78. Lu, Haitao; Que, Yumei; Wu, Xia; Guan, Tianbing; Guo, Hao. Metabolomics Deciphered Metabolic Reprogramming Required for Biofilm Formation. Scientific Reports (2019). https://doi.org/10.1038/s41598-019-49603-1
79. Howden, Benjamin P.; Giulieri, Stefano G.; Wong Fok Lung, Tania; Baines, Sarah L.; Sharkey, Liam K.; Lee, Jean Y. H.; Hachani, Abderrahman; Monk, Ian R.; Stinear, Timothy P. Staphylococcus aureus host interactions and adaptation. Nature Reviews Microbiology (2023). https://doi.org/10.1038/s41579-023-00852-y
80. Lin, Chih-Chun Janet; Wang, Meng C. Microbial metabolites regulate host lipid metabolism through NR5A–Hedgehog signalling. Nature Cell Biology (2017). https://doi.org/10.1038/ncb3515
81. Kuang, Huijuan; Dou, Geng; Cheng, Linfeng; Wang, Xiangdong; Xu, Haokun; Liu, Xuemei; Ding, Feng; Yang, Xiaoshan; Liu, Siying; Bao, Lili; Liu, Huan; Liu, Yao; Li, Bei; Jin, Yan; Liu, Shiyu Humoral regulation of iron metabolism by extracellular vesicles drives antibacterial response. Nature Metabolism (2023). https://doi.org/10.1038/s42255-022-00723-5
82. Rajeeve, Karthika; Vollmuth, Nadine; Janaki-Raman, Sudha; Wulff, Thomas F.; Baluapuri, Apoorva; Dejure, Francesca R.; Huber, Claudia; Fink, Julian; Schmalhofer, Maximilian; Schmitz, Werner; Sivadasan, Rajeeve; Eilers, Martin; Wolf, Elmar; Eisenreich, Wolfgang; Schulze, Almut; Seibel, Jürgen; Rudel, Thomas Reprogramming of host glutamine metabolism during Chlamydia trachomatis infection and its key role in peptidoglycan synthesis. Nature Microbiology (2020). https://doi.org/10.1038/s41564-020-0762-5
83. Wong Fok Lung, Tania; Monk, Ian R.; Acker, Karen P.; Mu, Andre; Wang, Nancy; Riquelme, Sebastián A.; Pires, Silvia; Noguera, Loreani P.; Dach, Felix; Gabryszewski, Stanislaw J.; Howden, Benjamin P.; Prince, Alice Staphylococcus aureus small colony variants impair host immunity by activating host cell glycolysis and inducing necroptosis. Nature Microbiology (2019). https://doi.org/10.1038/s41564-019-0597-0

Chapter 7
Impact of Antimicrobial Resistance on Virulence and Disease Spread

Vijay Soni and Aditya Upadhyay

7.1 Introduction by the Editor

Antimicrobial resistance (AMR) has emerged as one of the most pressing challenges in modern medicine, threatening the efficacy of treatments and the control of infectious diseases worldwide. While the direct consequences of AMR—such as treatment failure and increased mortality—are well recognized, its broader impact on microbial virulence, community dynamics, and disease spread is only beginning to be understood. This chapter explores the different ways in which resistance mechanisms intersect with microbial metabolism, host interactions, and ecological pressures, revealing a complex landscape where AMR not only enables survival but also reshapes pathogenic potential.

Microbial communities are inherently cooperative and metabolically interdependent. Recent studies show that auxotrophic organisms, unable to synthesize certain essential metabolites, are prevalent in host-associated microbiomes and contribute to drug tolerance. These auxotrophs enhance metabolite exchange within communities, increasing efflux activity and reducing intracellular drug concentrations. This metabolic interdependence fosters resilience against antimicrobial agents, suggesting that community-level interactions can amplify individual resistance traits and influence treatment outcomes.

Collateral sensitivity (CS), where resistance to one antibiotic increases susceptibility to another, offers a promising strategy to counteract AMR. In *Pseudomonas aeruginosa*, transient CS to tobramycin can be induced by dequalinium chloride, which overactivated efflux pumps. This approach temporarily sensitizes bacteria to

V. Soni (✉)
Department of Medicine, Weill Cornell Medicine, New York, NY, USA
e-mail: vis2032@med.cornell.edu

A. Upadhyay
Hospital for Special Surgery, New York, NY, USA

V. Soni, A. Upadhyay (eds.), *Metabolism of Bacterial Infection*,
https://doi.org/10.1007/978-3-031-46895-7_7

aminoglycosides without selecting for stable resistance mutations. Such findings highlight the therapeutic potential of exploiting metabolic and regulatory vulnerabilities in resistant pathogens.

Metabolomic profiling of volatile organic compounds (VOCs) provides another avenue for understanding and detecting AMR. VOC fingerprints can differentiate between antibiotic-sensitive and resistant strains of biothreat agents like Yersinia pestis, offering rapid diagnostic capabilities. Although challenges remain in translating these methods to in vivo settings, advances in thermal desorption-GC-MS techniques show promise for clinical applications.

The environmental consequences of mass drug administration (MDA) programs are also significant. Widespread use of antibiotics like azithromycin in public health campaigns can lead to the release of unmetabolized drugs into ecosystems, promoting the emergence of resistant strains in environmental microbiota. Studies underscore the need for integrated surveillance systems to monitor antimicrobial residues and assess ecological risks, balancing public health benefits with environmental stewardship.

At the molecular level, metabolic genes play a critical role in modulating resistance. In *Neisseria gonorrhoeae*, a mutation in the β-carbonic anhydrase gene CanB compensates for fitness costs associated with ciprofloxacin resistance, facilitating its acquisition. This metabolic adaptation illustrates how resistance evolution is shaped not only by drug pressure but also by the physiological context of the pathogen.

Temperate bacteriophages also influence resistance and virulence. Integration of Shiga toxin-encoding phage ϕ24B into *Escherichia coli* alters host metabolism, enhancing tolerance to antimicrobials and utilization of alternative nutrients. These findings suggest that phage-mediated metabolic reprogramming can confer selective advantages beyond toxin production, contributing to persistence and spread.

Predictive models based on metabolic fitness offer new tools for anticipating resistance evolution. In *E. coli*, resistance to streptomycin evolves through membrane-related mutations that reduce drug uptake. These models link resistance levels to growth rates and can forecast evolutionary trajectories under different environmental conditions, aiding in the design of more effective treatment protocols.

Finally, conjugative plasmids carrying both metabolic and resistance genes are highly prevalent in *E. coli*. These plasmids facilitate horizontal gene transfer and provide metabolic advantages that enhance survival under antibiotic pressure. Understanding the co-selection of metabolic and resistance traits is essential for developing strategies to curb the spread of AMR.

Together, these studies reveal that antimicrobial resistance is not an isolated phenomenon but a dynamic force that reshapes microbial physiology, community structure, and ecological interactions. This chapter provides a comprehensive overview of how resistance mechanisms intersect with metabolism and virulence, offering insights into the evolution and dissemination of AMR and informing future therapeutic and public health strategies.

7.2 Machine-Generated Summaries

Disclaimer: The summaries in this chapter were generated from Springer Nature publications using abstractive AI auto-summarization: An abstraction-based summarizer creates new text based on deep learning. New phrases are created to summarize the content. As the constituted sentences are machine selected, they may not fully reflect the body of the work, so we strongly advise that the original content is read and cited. The auto-generated summaries were curated by the editor to meet Springer Nature publication standards.

Microbial Communities Form Rich Extracellular Metabolomes That Foster Metabolic Interactions and Promote Drug Tolerance [56]

This is a machine-generated summary of:

Yu, Jason S. L.; Correia-Melo, Clara; Zorrilla, Francisco; Herrera-Dominguez, Lucia; Wu, Mary Y.; Hartl, Johannes; Campbell, Kate; Blasche, Sonja; Kreidl, Marco; Egger, Anna-Sophia; Messner, Christoph B.; Demichev, Vadim; Freiwald, Anja; Mülleder, Michael; Howell, Michael; Berman, Judith; Patil, Kiran R.; Alam, Mohammad Tauqeer; Ralser, Markus Microbial Communities Form Rich Extracellular Metabolomes That Foster Metabolic Interactions and Promote Drug Tolerance [56]

Published in: Nature Microbiology (2022)

Link to original: https://doi.org/10.1038/s41564-022-01072-5

If you want to cite the papers, please refer to the original.

For technical reasons we could not place the page where the original quote is coming from.

Research Highlights

Microbial communities are complex systems where the exchange of metabolites plays a crucial role in their physiology and survival. Metabolite exchange, particularly in microbial communities, is essential for cross-feeding and collective survival, allowing cells to save resources and improve biomass formation. Auxotrophs, which lack the ability to synthesize certain essential metabolites, rely on the extracellular availability of these compounds for growth. This dependency highlights a knowledge gap regarding how auxotrophs persist in communities without being at a disadvantage. Now, a new study by Jason S. L. Yu and colleagues explores the prevalence of auxotrophs in microbial communities and their impact on drug resilience.

The study utilized data from the Earth Microbiome Project (EMP) to analyze the frequency of auxotrophs in both free-living and host-associated communities. The researchers found that auxotrophs are prevalent across various environments, with a particularly high abundance in host-associated communities. To investigate the physiological consequences of this prevalence, they examined growth data from gut microbial species exposed to bioactive drugs. The study revealed that amino acid auxotrophs generally exhibited better growth in the presence of drugs compared to prototrophs. To further understand this phenomenon, the researchers employed a yeast model system called SeMeCos, which allowed them to replicate and study the increased drug resilience observed in auxotrophs. The model demonstrated that auxotrophs enhance metabolite exchange interactions within communities, leading to increased efflux activity and reduced intracellular drug concentrations.

The findings suggest that auxotrophy contributes to increased drug resilience through enhanced metabolite exchange interactions within microbial communities. This study provides insights into how metabolic interactions can influence community dynamics and drug responses. Future studies could explore the evolutionary implications of these interactions and their role in the spread of drug resistance genes. Additionally, understanding the mechanisms underlying increased efflux activity could inform strategies for predicting drug efficacy and managing antimicrobial resistance in complex microbial ecosystems.

Tackling Antibiotic Resistance by Inducing Transient and Robust Collateral Sensitivity [57]

This is a machine-generated summary of:

Hernando-Amado, Sara; Laborda, Pablo; Martínez, José Luis Tackling Antibiotic Resistance by Inducing Transient and Robust Collateral Sensitivity [57]

Published in: Nature Communications (2023)

Link to original: https://doi.org/10.1038/s41467-023-37357-4

If you want to cite the papers, please refer to the original.

For technical reasons we could not place the page where the original quote is coming from.

Research Highlights

Antibiotic resistance in bacterial pathogens poses a significant threat to human health, with *Pseudomonas aeruginosa* being a major concern due to its role in nosocomial and chronic infections. This bacterium exhibits low intrinsic susceptibility to antibiotics and can enhance its resistance through mutations, particularly involving multidrug efflux pumps like MexCD-OprJ. These pumps are often overexpressed due to mutations in regulatory genes or exposure to specific inducers, complicating treatment strategies. Now, a new study by Sara Hernando-Amado and colleagues explores the potential of exploiting collateral sensitivity (CS) in *P. aeruginosa* to improve antibiotic efficacy.

The study investigates whether transient CS to tobramycin can be induced by dequalinium chloride (DC), an antiseptic known to trigger MexCD-OprJ overexpression, thereby temporarily increasing ciprofloxacin resistance. The researchers conducted experiments on various *P. aeruginosa* strains, including clinical isolates, to assess changes in antibiotic susceptibility when exposed to DC. They measured minimal inhibitory concentrations (MICs) of ciprofloxacin and tobramycin with and without DC presence. The results showed that DC effectively induced a transient hyper-susceptible state to tobramycin across different genetic backgrounds without leading to stable resistance mutations.

The findings suggest that using DC could be a viable strategy for enhancing the effectiveness of aminoglycosides like tobramycin against resistant *P. aeruginosa* strains. Future studies could focus on optimizing this approach for clinical applications, potentially incorporating DC into treatment regimens for pulmonary infections caused by *P. aeruginosa*. This research highlights the potential of leveraging transient CS as an innovative therapeutic strategy, offering a promising avenue for tackling antibiotic-resistant infections while minimizing the risk of selecting for stable resistant mutants.

VOC Fingerprints: Metabolomic Signatures of Biothreat Agents with and Without Antibiotic Resistance [58]

This is a machine-generated summary of:

Dailey, Allyson; Saha, Jessica; Zaidi, Fatima; Abdirahman, Hafsa; Haymond, Amanda; Alem, Farhang; Hakami, Ramin; Couch, Robin VOC Fingerprints: Metabolomic Signatures of Biothreat Agents with and Without Antibiotic Resistance [58]

Published in: Scientific Reports (2020)

Link to original: https://doi.org/10.1038/s41598-020-68622-x

If you want to cite the papers, please refer to the original.

For technical reasons we could not place the page where the original quote is coming from.

Research Highlights

The threat posed by Category A and B biothreat agents, such as certain bacteria, viruses, and toxins, is significant due to their potential for widespread dissemination and high morbidity/mortality rates. Traditional detection methods like polymerase chain reaction and enzyme-linked immunosorbent assays, while effective, are often time-consuming and costly. A promising alternative is the use of volatile organic compounds (VOCs) for rapid diagnostics. Now, a new study by Allyson Dailey and colleagues explores the potential of microbial VOC metabolomic profiling to differentiate between liquid cultures of *Francisella tularensis*, *Burkholderia pseudomallei*, and *Yersinia pestis*.

The study employed multifiber headspace solid phase microextraction (hSPME) combined with gas chromatography-flame ionization detection (GC-FID) to derive unique mVOC fingerprints for these bacteria. The researchers also tested the influence of environmental factors on these profiles and developed a device for simultaneous multifiber hSPME to reduce analysis time. While 'simultaneous multifiber headspace solid-phase microextraction' (simultihSPME) effectively differentiated antibiotic-sensitive and resistant strains in vitro, it failed to distinguish infected from healthy mice in an animal model. However, thermal desorption coupled with GC-MS successfully differentiated VOC profiles from healthy mice versus those infected with either antibiotic-sensitive or resistant strains of *Y. pestis*.

The study advances the field by demonstrating that mVOC profiling can uniquely identify bacterial strains under various conditions and differentiate antibiotic-resistant strains. Future studies could explore whether this method can discern other forms of antimicrobial resistance. Although simulti-hSPME was less effective for rapid in vivo diagnostics, thermal desorption-GC-MS shows promise for clinical applications by rapidly distinguishing infected from healthy subjects. This approach could be pivotal in addressing future biothreats through enhanced diagnostic capabilities in both civilian and military contexts.

Environmental Impacts of Mass Drug Administration Programs: Exposures, Risks, and Mitigation of Antimicrobial Resistance [59]

This is a machine-generated summary of:

Konopka, Joanna K.; Chatterjee, Pranab; LaMontagne, Connor; Brown, Joe Environmental Impacts of Mass Drug Administration Programs: Exposures, Risks, and Mitigation of Antimicrobial Resistance [59]

Published in: Infectious Diseases of Poverty (2022)

Link to original: https://doi.org/10.1186/s40249-022-01000-z

If you want to cite the papers, please refer to the original.

For technical reasons we could not place the page where the original quote is coming from.

Research Highlights

Mass drug administration (MDA) programs have been instrumental in combating neglected tropical diseases (NTDs) by providing widespread treatment without the need for individual diagnosis. While these programs have successfully reduced childhood morbidity and mortality, they also raise concerns about potential environmental impacts, particularly the exacerbation of antimicrobial resistance (AMR). The use of antimicrobial agents in MDA can lead to significant quantities of unmetabolized drugs entering the environment, potentially contributing to AMR. Despite these concerns, environmental monitoring to assess AMR following MDA has been largely overlooked. Now, a new study by Joanna K. Konopka and colleagues reviews the current understanding of how MDA programs introduce antimicrobials into the environment and their potential ecological impacts.

The study conducted a comprehensive scoping review using databases such as PubMed, Web of Science, and Google Scholar to gather relevant literature on MDA and its environmental consequences. The authors focused on azithromycin, a commonly used antibiotic in MDA programs for trachoma and yaws control, tracing its journey from human excretion to environmental exposure. They highlighted how azithromycin-resistant strains of bacteria like *Escherichia coli* and *Streptococcus pneumoniae* have been detected in communities following MDA. The review also explored the persistence of antimicrobial residues in the environment and their potential to select for resistant genes in microbial communities. The authors consulted with experts in environmental microbiology and AMR to ensure a thorough evaluation of the available evidence.

The findings underscore the need for integrated monitoring systems to assess AMR emergence alongside MDA initiatives. Future studies should focus on understanding the long-term ecological impacts of antimicrobial residues and developing strategies to mitigate these effects. The authors recommend implementing environmental risk assessments and surveillance systems to monitor antimicrobial concentrations in waste streams and ecosystems around treated populations. By doing so, stakeholders can better understand the unintended consequences of MDA programs and work toward minimizing their ecological footprint while maintaining public health benefits.

CanB Is a Metabolic Mediator of Antibiotic Resistance in Neisseria gonorrhoeae *[60]*

This is a machine-generated summary of:

Rubin, Daniel H. F.; Ma, Kevin C.; Westervelt, Kathleen A.; Hullahalli, Karthik; Waldor, Matthew K.; Grad, Yonatan H. CanB Is a Metabolic Mediator of Antibiotic Resistance in *Neisseria gonorrhoeae* [60]

Published in: Nature Microbiology (2023)
Link to original: https://doi.org/10.1038/s41564-022-01282-x

If you want to cite the papers, please refer to the original.

For technical reasons we could not place the page where the original quote is coming from.

Research Highlights

Neisseria gonorrhoeae, the bacterium responsible for the widespread sexually transmitted infection gonorrhea, has developed resistance to all first-line antibiotics, including ciprofloxacin and ceftriaxone. This resistance varies across different strains, with two major lineages identified: lineage A, associated with men who have sex with men and high antibiotic resistance, and lineage B, linked to heterosexuals with lower resistance levels. The reasons behind these variations in resistance are not fully understood. Now, a new study by Daniel H. F. Rubin and colleagues explores the role of metabolic factors in antibiotic resistance acquisition in *N. gonorrhoeae*.

The study employed a genome-wide association study (GWAS) to investigate CO2 dependence in *N. gonorrhoeae*, revealing that a single mutation in the β-carbonic anhydrase gene CanB is necessary and sufficient for this trait. The researchers assessed clinical and laboratory strains for growth with and without supplemental CO2, identifying a significant association with a missense mutation in CanB. Further experiments demonstrated that this mutation compensates for fitness costs associated with gyrase mutations conferring ciprofloxacin resistance. The study also found that CanB variants affect intracellular pH buffering and growth under anaerobic conditions, suggesting an adaptation to specific environmental niches.

The findings highlight how metabolic variation can influence antibiotic resistance acquisition in *N. gonorrhoeae*. The CanB variant facilitates the acquisition of ciprofloxacin resistance by compensating for fitness costs related to gyrase mutations, without directly affecting drug susceptibility or tolerance. Future studies could explore how gyrase mutations alter the metabolic landscape of *N. gonorrhoeae* cells and further investigate the complex interactions between metabolic genes and antibiotic resistance determinants. This research provides new insights into the evolution of antibiotic resistance and underscores the importance of considering metabolic factors in understanding bacterial adaptation to antimicrobial pressures.

Shigatoxin Encoding Bacteriophage ϕ24$_B$ Modulates Bacterial Metabolism to Raise Antimicrobial Tolerance [61]

This is a machine-generated summary of:

Holt, G. S.; Lodge, J. K.; McCarthy, A. J.; Graham, A. K.; Young, G.; Bridge, S. H.; Brown, A. K.; Veses-Garcia, M.; Lanyon, C. V.; Sails, A.; Allison, H. E.; Smith, D. L. Shigatoxin Encoding Bacteriophage ϕ24$_B$ Modulates Bacterial Metabolism to Raise Antimicrobial Tolerance [61]

Published in: Scientific Reports (2017)

Link to original: https://doi.org/10.1038/srep40424

Research Highlights

Shiga toxin-encoding *Escherichia coli* (STEC) is a significant public health concern due to its potential to cause severe gastrointestinal infections in humans, ranging from bloody diarrhea to life-threatening conditions like hemolytic uremic syndrome. The pathogenicity of STEC is largely attributed to Shiga toxins, which are encoded by temperate bacteriophages (Stx-phages) that integrate into the bacterial genome. Despite the known role of Stx-phages in enhancing bacterial virulence, the specific mechanisms by which these phages influence bacterial physiology and antimicrobial tolerance remain poorly understood. Now, a new study by G. S. Holt and colleagues investigates how the integration of Stx-phage B into *E. coli* affects bacterial metabolism and antimicrobial tolerance.

Holt et al. employed a combination of phenotypic assays and untargeted metabolomics to explore the impact of Stx-phage B integration on *E. coli* physiology. The study utilized Biolog Phenotype MicroArrays to assess changes in respiration and metabolism, revealing that lysogeny with phage B enhances *E. coli*'s ability to utilize alternative phosphate sources and increases tolerance to various antimicrobials, including chloroxylenol and 8-hydroxyquinoline. Additionally, high-resolution liquid chromatography-mass spectrometry (LC-MS) was used to profile metabolic changes, identifying significant alterations in metabolites related to biotin and fatty acid synthesis. These findings suggest that phage integration modulates host metabolism, potentially conferring a selective advantage under antimicrobial pressure.

The study advances our understanding of phage–host interactions by demonstrating that Stx-phage B integration can lead to metabolic shifts that enhance antimicrobial tolerance in *E. coli*. This research highlights the potential for temperate phages to influence bacterial adaptation beyond traditional antibiotic resistance mechanisms, suggesting a broader role for phages in microbial ecology and evolution. Future studies could further elucidate the specific genetic pathways involved in these metabolic changes and explore their implications for managing STEC infections in agricultural settings where antimicrobials are commonly used.

Metabolic Fitness Landscapes Predict the Evolution of Antibiotic Resistance [62]

This is a machine-generated summary of:

Pinheiro, Fernanda; Warsi, Omar; Andersson, Dan I.; Lässig, Michael Metabolic Fitness Landscapes Predict the Evolution of Antibiotic Resistance [62]

Published in: Nature Ecology & Evolution (2021)

Link to original: https://doi.org/10.1038/s41559-021-01397-0

If you want to cite the papers, please refer to the original.

For technical reasons we could not place the page where the original quote is coming from.

Research Highlights

Modern human and veterinary medicine heavily depend on antibiotics to treat bacterial infections. However, the overuse and misuse of these drugs have led to a global rise in antibiotic-resistant bacteria, reducing treatment efficacy and increasing health risks. Understanding the conditions under which resistance evolves and predicting its dynamics are crucial for epidemiology, clinical microbiology, and evolutionary biology. Now, a new study by Fernanda Pinheiro and colleagues explores the evolution of antibiotic resistance in *Escherichia coli*, focusing on metabolic fitness models that predict resistance trajectories under varying drug and nutrient conditions.

The study investigates resistance evolution in response to the aminoglycoside streptomycin, which inhibits translation by binding to ribosomes. The researchers conducted selection experiments with *E. coli* in different drug and nutrient environments, using both rich lysogeny broth and minimal glycerol media. They applied streptomycin concentrations slightly above the minimum inhibitory concentration to elicit resistance mutations. Eight mutants were randomly selected from each culture for whole-genome sequencing and growth rate analysis across various drug levels. The study found that resistance mutations were distributed across the *E. coli* genome but showed consistent patterns in functional targets and growth characteristics, particularly affecting membrane-related functions linked to the electron transport chain.

The findings reveal a universal mechanism of membrane-based resistance evolution, where mutations reduce drug uptake by altering membrane permeability. This mechanism links two independent phenotypes: resistance level and drug-free growth rate. The study's metabolic fitness models successfully predict evolutionary trajectories of dosage-dependent resistance levels and growth rates, including previously unseen mutations. Future studies could expand these models to include multiple mutations and compensatory traits, enhancing their predictive power for clinical applications. By identifying probable resistance pathways, this research aids in designing effective drugs and treatment protocols against antibiotic-resistant bacteria.

Metabolic Genes on Conjugative Plasmids Are Highly Prevalent in Escherichia coli *and Can Protect Against Antibiotic Treatment [63]*

This is a machine-generated summary of:

Palomino, Alana; Gewurz, Danya; DeVine, Lela; Zajmi, Ujana; Moralez, Jenifer; Abu-Rumman, Fatima; Smith, Robert P.; Lopatkin, Allison J. Metabolic Genes on Conjugative Plasmids Are Highly Prevalent in *Escherichia coli* and Can Protect Against Antibiotic Treatment [63]

Published in: The ISME Journal: Multidisciplinary Journal of Microbial Ecology (2022)

Link to original: https://doi.org/10.1038/s41396-022-01329-1

If you want to cite the papers, please refer to the original.

For technical reasons we could not place the page where the original quote is coming from.

Research Highlights

Horizontal gene transfer, particularly through plasmid conjugation, is a crucial mechanism in microbial ecology and evolution, allowing bacteria to rapidly adapt to environmental changes. Plasmids can transfer genes across diverse bacterial species, including those conferring antibiotic resistance, which is a significant public health concern. While the role of plasmids in spreading antibiotic resistance is well-documented, the potential contribution of plasmid-encoded metabolic genes to antibiotic resistance remains underexplored. Now, a new study by Alana Palomino and colleagues investigates the prevalence and impact of metabolic genes on transferrable plasmids in *Escherichia coli*, revealing their potential role in antibiotic resistance.

The study involved curating a comprehensive dataset of *E. coli* plasmids from publicly available genomes to assess the prevalence and distribution of metabolic and antibiotic resistance genes. Using bioinformatics tools like MOB-suite for plasmid transferability prediction and Prokka for gene annotation, the researchers identified significant co-occurrence patterns between metabolic and antibiotic resistance genes on these plasmids. They conducted experiments to evaluate the protective effects of select metabolic genes against antibiotics by expressing them in *E. coli* strains and measuring changes in antibiotic susceptibility. The findings showed that certain metabolic genes could confer resistance to antibiotics, suggesting that these genes might play a role in bacterial survival under antibiotic pressure.

The study advances our understanding of the complex interactions between metabolic and antibiotic resistance genes on plasmids. It highlights that metabolic genes may contribute to antibiotic resistance dissemination by providing additional survival advantages under antibiotic treatment. Future studies could explore the ecological and evolutionary dynamics influencing these gene associations and investigate their roles in other bacterial species beyond *E. coli*. Understanding these interactions could inform strategies to mitigate the spread of antibiotic resistance by targeting both canonical resistance mechanisms and associated metabolic pathways.

7.3 Conclusion

The studies presented in this chapter illuminate the evolving relationship between antimicrobial resistance (AMR), microbial metabolism, and disease dynamics. Far beyond its role in conferring survival under drug pressure, AMR is shown to influence microbial physiology, community behavior, virulence, and ecological impact. These findings underscore the need to view resistance not as a static trait but as a dynamic and integrative phenomenon that reshapes the biology of infection and the strategies required to combat it.

One of the most compelling insights is the role of metabolic interdependence in promoting drug tolerance. Auxotrophs, traditionally viewed as metabolically disadvantaged, are revealed to be central players in microbial communities, particularly in host-associated environments. Their reliance on extracellular metabolites fosters enhanced exchange interactions, which in turn increase efflux activity and reduce intracellular drug accumulation. This community-level resilience challenges conventional approaches to antimicrobial therapy and suggests that targeting metabolic networks may be as important as inhibiting resistance genes.

The concept of collateral sensitivity (CS) offers a novel therapeutic strategy to exploit the vulnerabilities of resistant pathogens. In *Pseudomonas aeruginosa*, transient CS induced by dequalinium chloride sensitizes bacteria to aminoglycosides without selecting for stable resistance mutations. This approach demonstrates that resistance can be strategically manipulated through metabolic and regulatory pathways, opening new avenues for combination therapies that minimize resistance evolution.

Metabolomic profiling, particularly of volatile organic compounds (VOCs), provides a powerful diagnostic tool for identifying resistant strains and biothreat agents. Although challenges remain in translating these techniques to in vivo settings, advances in thermal desorption-GC-MS show promise for rapid and noninvasive detection. These methods could revolutionize clinical diagnostics and biodefense by enabling early identification of resistant infections.

Environmental considerations are increasingly critical in the context of AMR. Mass drug administration (MDA) programs, while effective in controlling neglected tropical diseases, inadvertently contribute to environmental antimicrobial contamination. The persistence of drug residues and the emergence of resistant strains in treated communities highlight the need for integrated surveillance systems and environmental risk assessments. Balancing public health benefits with ecological sustainability is essential for long-term AMR mitigation.

At the molecular level, metabolic adaptations play a pivotal role in resistance evolution. In *Neisseria gonorrhoeae*, a mutation in the CanB gene compensates for fitness costs associated with ciprofloxacin resistance, facilitating its acquisition. This finding illustrates how metabolic context can influence the trajectory of resistance development, emphasizing the importance of considering physiological factors in resistance studies.

Phage-mediated modulation of bacterial metabolism adds another layer of complexity. The integration of Shiga toxin-encoding phage ϕ24B into *E. coli* enhances antimicrobial tolerance and nutrient utilization, suggesting that temperate phages contribute to resistance and virulence through metabolic reprogramming. These interactions highlight the ecological and evolutionary significance of phage–host dynamics in shaping microbial fitness.

Predictive models based on metabolic fitness landscapes offer valuable tools for anticipating resistance evolution. In *E. coli*, resistance to streptomycin evolves through membrane-related mutations that reduce drug uptake. These models successfully link resistance levels to growth rates and forecast evolutionary trajectories, providing a framework for designing more effective treatment protocols and anticipating resistance pathways.

Finally, the prevalence of metabolic genes on conjugative plasmids in *E. coli* reveals a hidden dimension of resistance dissemination. These genes not only confer metabolic advantages but also enhance survival under antibiotic pressure, suggesting that plasmid-mediated co-selection of resistance and metabolic traits is a significant driver of AMR spread. Understanding these interactions is crucial for developing strategies to disrupt horizontal gene transfer and limit resistance propagation.

In conclusion, antimicrobial resistance is a multifaceted phenomenon that intersects with microbial metabolism, community ecology, and host–pathogen interactions. The studies in this chapter demonstrate that resistance mechanisms are deeply embedded in the physiological and ecological fabric of microbial life. Addressing AMR requires a holistic approach that integrates molecular biology, systems ecology, clinical medicine, and environmental science. By expanding our understanding of how resistance influences and is influenced by metabolic and ecological factors, we can develop more nuanced and effective strategies to combat the global threat of antimicrobial resistance.

Bibliography

1. Mee, M. T., Collins, J. J., Church, G. M. & Wang, H. H. Syntrophic exchange in synthetic microbial communities. *Proc. Natl Acad. Sci. USA 111*, E2149–E2156 (2014).
2. Wang, Y.-P., Li, J.-T., Qu, J., Yin, M. & Lei, Q.-Y. Metabolite sensing and signaling in cancer. *J. Biol. Chem. 295*, 11938–11946 (2020).
3. Pinu, F. R. et al. Metabolite secretion in microorganisms: the theory of metabolic overflow put to the test. *Metabolomics 14*, 43 (2018).
4. Douglas, A. E. The microbial exometabolome: ecological resource and architect of microbial communities. *Philos. Trans. R. Soc. Lond. B Biol. Sci. 375*, 20190250 (2020).
5. Blasche, S. et al. Metabolic cooperation and spatiotemporal niche partitioning in a kefir microbial community. *Nat. Microbiol. 6*, 196–208 (2021).
6. D'Souza, G. et al. Ecology and evolution of metabolic cross-feeding interactions in bacteria. *Nat. Prod. Rep. 35*, 455–488 (2018).
7. Maier, L. et al. Extensive impact of non-antibiotic drugs on human gut bacteria. *Nature 555*, 623–628 (2018).
8. Lee, S., Lim, W. A. & Thorn, K. S. Improved blue, green, and red fluorescent protein tagging vectors for *S. cerevisiae*. *PLoS ONE 8*, e67902 (2013).
9. Vowinckel, J., Hartl, J., Butler, R. & Ralser, M. MitoLoc: a method for the simultaneous quantification mitochondrial network morphology and membrane potential in single cells. *Mitochondrion 24*, 77–86 (2015).
10. Szybalski, W. & Bryson, V. Genetic studies on microbial cross resistance to toxic agents. I. Cross resistance of Escherichia coli to fifteen antibiotics. *J. Bacteriol. 64*, 489–499 (1952).
11. Pal, C., Papp, B. & Lazar, V. Collateral sensitivity of antibiotic-resistant microbes. *Trends Microbiol. 23*, 401–407 (2015).
12. Lopez-Causape, C., Cabot, G., Del Barrio-Tofino, E. & Oliver, A. The versatile mutational resistome of *Pseudomonas aeruginosa*. *Front. Microbiol. 9*, 685 (2018).
13. Lopez-Causape, C. et al. Evolution of the *Pseudomonas aeruginosa* mutational resistome in an international Cystic Fibrosis clone. *Sci. Rep. 7*, 5555 (2017).
14. Armstrong, D. et al. Evidence for spread of a clonal strain of *Pseudomonas aeruginosa* among cystic fibrosis clinics. *J. Clin. Microbiol. 41*, 2266–2267 (2003).
15. Knopp, M. & Andersson, D. I. Predictable phenotypes of antibiotic resistance mutations. *mBio 9*, https://doi.org/10.1128/mBio.00770-18 (2018).

16. Hernando-Amado, S., Laborda, P., Valverde, J. R. & Martínez, J. L. Mutational background influences P. aeruginosa ciprofloxacin resistance evolution but preserves collateral sensitivity robustness. *Proc. Natl Acad. Sci. USA 119*, e2109370119 (2022).
17. Laborda, P., Alcalde-Rico, M., Blanco, P., Martinez, J. L. & Hernando-Amado, S. Novel inducers of the expression of multidrug efflux pumps that trigger *Pseudomonas aeruginosa* transient antibiotic resistance. *Antimicrob. Agents Chemother. 63*, e01095–19 (2019).
18. Hernando-Amado, S. et al. Rapid phenotypic convergence towards collateral sensitivity in clinical isolates of Pseudomonas aeruginosa presenting different genomic backgrounds. *Microbiol. Spectr. 11*, e0227622 (2022).
19. Mulet, X. et al. Antagonistic interactions of *Pseudomonas aeruginosa* antibiotic resistance mechanisms in planktonic but not biofilm growth. *Antimicrob. Agents Chemother. 55*, 4560–4568 (2011).
20. Rees, C. A. *et al.* Detection of high-risk carbapenem-resistant *Klebsiella pneumoniae* and *Enterobacter cloacae* isolates using volatile molecular profiles. *Sci. Rep. 8*, 13297 (2018).
21. Maier, T. M. *et al.* Construction and characterization of a highly efficient *Francisella* shuttle plasmid. *Appl. Environ. Microbiol. 70*, 7511–7519 (2004).
22. Smits HL. Prospects for the control of neglected tropical diseases by mass drug administration. Expert Rev Anti Infect Ther. 2009;7:37–56.
23. Baquero F, Martínez J-L, Cantón R. Antibiotics and antibiotic resistance in water environments. Curr Opin Biotechnol. 2008;19:260–5.
24. Wellington EM, Boxall AB, Cross P, Feil EJ, Gaze WH, Hawkey PM, et al. The role of the natural environment in the emergence of antibiotic resistance in Gram-negative bacteria. Lancet Infect Dis. 2013;13:155–65.
25. Skalet AH, Cevallos V, Ayele B, Gebre T, Zhou Z, Jorgensen JH, et al. Antibiotic selection pressure and macrolide resistance in nasopharyngeal *Streptococcus pneumoniae*: a cluster-randomized clinical trial. PLoS Med. 2010;7: e1000377.
26. Leach AJ, Shelby-James TM, Mayo M, Gratten M, Laming AC, Currie BJ, et al. A prospective study of the impact of community-based azithromycin treatment of trachoma on carriage and resistance of *Streptococcus pneumoniae*. Clin Infect Dis. 1997;24:356–62.
27. Seidman JC, Coles CL, Silbergeld EK, Levens J, Mkocha H, Johnson LB, et al. Increased carriage of macrolide-resistant fecal *E. coli* following mass distribution of azithromycin for trachoma control. Int J Epidemiol. 2014;43:1105–13.
28. Seidman JC, Johnson LB, Levens J, Mkocha H, Muñoz B, Silbergeld EK, et al. Longitudinal comparison of antibiotic resistance in diarrheagenic and non-pathogenic escherichia coli from young Tanzanian children. Front Microbiol. 2016;7:1420.
29. O'Brien KS, Emerson P, Hooper PJ, Reingold AL, Dennis EG, Keenan JD, et al. Antimicrobial resistance following mass azithromycin distribution for trachoma: a systematic review. Lancet Infect Dis. 2019;19:e14–25.
30. Doan T, Arzika AM, Hinterwirth A, Maliki R, Zhong L, Cummings S, et al. Macrolide resistance in MORDOR I—a cluster-randomized trial in niger. N Engl J Med. 2019;380:2271–3.
31. Moser W, Schindler C, Keiser J. Efficacy of recommended drugs against soil transmitted helminths: systematic review and network meta-analysis. BMJ. 2017;358: j4307.
32. Hanna N, Sun P, Sun Q, Li X, Yang X, Ji X, et al. Presence of antibiotic residues in various environmental compartments of Shandong province in eastern China: Its potential for resistance development and ecological and human risk. Environ Int. 2018;114:131–42.
33. Bengtsson-Palme J, Larsson DGJ. Concentrations of antibiotics predicted to select for resistant bacteria: Proposed limits for environmental regulation. Environ Int. 2016;86:140–9.
34. Golparian, D. et al. Genomic evolution of *Neisseria gonorrhoeae* since the preantibiotic era (1928-2013): antimicrobial use/misuse selects for resistance and drives evolution. *BMC Genomics 21*, 116 (2020).
35. Gransden, W. R., Warren, C. A., Phillips, I., Hodges, M. & Barlow, D. Decreased susceptibility of *Neisseria gonorrhoeae* to ciprofloxacin. *Lancet (Lond., Engl.) 335*, 51 (1990).

36. Unemo, M. & Shafer, W. M. Antimicrobial resistance in *Neisseria gonorrhoeae* in the 21st century: past, evolution, and future. *Clin. Microbiol. Rev. 27*, 587–613 (2014).
37. *Sexually Transmitted Disease Surveillance – 2020* (CDC, 2020).
38. WHO Gonococcal AMR Surveillance Programme. *WHO* https://www.who.int/data/gho/data/themes/topics/who-gonococcal-amr-surveillance-programme-who-gasp (2022).
39. Rubin, D. H. F., Ross, J. D. C. & Grad, Y. H. The frontiers of addressing antibiotic resistance in *Neisseria gonorrhoeae. Transl. Res.* https://doi.org/10.1016/j.trsl.2020.02.002 (2020).
40. Ito, M. et al. Emergence and spread of *Neisseria gonorrhoeae* clinical isolates harboring mosaic-like structure of penicillin-binding protein 2 in Central Japan. *Antimicrobial Agents Chemother. 49*, 137–143 (2005).
41. Lopatkin, A. J. et al. Clinically relevant mutations in core metabolic genes confer antibiotic resistance. *Science 371*, eaba0862 (2021).
42. Pinheiro, F., Warsi, O., Andersson, D. I. & Lässig, M. Metabolic fitness landscapes predict the evolution of antibiotic resistance. *Nat. Ecol. Evol. 5*, 677–687 (2021).
43. Andersson, D. I. & Hughes, D. Antibiotic resistance and its cost: is it possible to reverse resistance? *Nat. Rev. Microbiol. 8*, 260–271 (2010).
44. MacLean, R. C. & San Millan, A. The evolution of antibiotic resistance. *Sci. (N. Y., N. Y.) 365*, 1082–1083 (2019).
45. Sousa, A., Magalhães, S. & Gordo, I. Cost of antibiotic resistance and the geometry of adaptation. *Mol. Biol. Evol. 29*, 1417–1428 (2012).
46. Sommer, M. O. A., Munck, C., Toft-Kehler, R. V. & Andersson, D. I. Prediction of antibiotic resistance: time for a new preclinical paradigm? *Nat. Rev. Microbiol. 15*, 689–696 (2017).
47. Vincent, L. R. et al. In vivo-selected compensatory mutations restore the fitness cost of mosaic pena alleles that confer ceftriaxone resistance in *Neisseria gonorrhoeae. mBio 9*, e01905–e01917 (2018).
48. Smith, D. L. et al. Comparative genomics of Shiga toxin encoding bacteriophages. *BMC genomics 13*, 311 (2012).
49. Unkmeir, A. & Schmidt, H. Structural analysis of phage-borne stx genes and their flanking sequences in Shiga toxin-producing Escherichia coli and Shigella dysenteriae type 1 strains. *Infection and immunity 68*, 4856–64 (2000).
50. Łuksza, M. & Lässig, M. A predictive fitness model for influenza. *Nature 507*, 57–61 (2014).
51. Lässig, M., Mustonen, V. & Walczak, A. M. Predicting evolution. *Nat. Ecol. Evol. 1*, 0077 (2017).
52. Greulich, P., Scott, M., Evans, M. R. & Allen, R. J. Growth-dependent bacterial susceptibility to ribosome-targeting antibiotics. *Mol. Syst. Biol. 11*, 796 (2015).
53. Lobritz MA, Belenky P, Porter CBM, Gutierrez A, Yang JH, Schwarz EG, et al. Antibiotic efficacy is linked to bacterial cellular respiration. Proc Natl Acad Sci USA. 2015;112:8173–80.
54. Brynildsen MP, Winkler JA, Spina CS, MacDonald IC, Collins JJ. Potentiating antibacterial activity by predictably enhancing endogenous microbial ROS production. Nat Biotechnol. 2013;31:160–5.
55. Kohanski MA, Dwyer DJ, Hayete B, Lawrence CA, Collins JJ. A Common mechanism of cellular death induced by bactericidal antibiotics. Cell 2007;130:797–810.
56. Yu, Jason S. L.; Correia-Melo, Clara; Zorrilla, Francisco; Herrera-Dominguez, Lucia; Wu, Mary Y.; Hartl, Johannes; Campbell, Kate; Blasche, Sonja; Kreidl, Marco; Egger, Anna-Sophia; Messner, Christoph B.; Demichev, Vadim; Freiwald, Anja; Mülleder, Michael; Howell, Michael; Berman, Judith; Patil, Kiran R.; Alam, Mohammad Tauqeer; Ralser, Markus. Microbial communities form rich extracellular metabolomes that foster metabolic interactions and promote drug tolerance. Nature Microbiology (2022). https://doi.org/10.1038/s41564-022-01072-5
57. Hernando-Amado, Sara; Laborda, Pablo; Martínez, José Luis. Tackling antibiotic resistance by inducing transient and robust collateral sensitivity. Nature Communications (2023). https://doi.org/10.1038/s41467-023-37357-4

58. Dailey, Allyson; Saha, Jessica; Zaidi, Fatima; Abdirahman, Hafsa; Haymond, Amanda; Alem, Farhang; Hakami, Ramin; Couch, Robin. VOC fingerprints: metabolomic signatures of biothreat agents with and without antibiotic resistance. Scientific Reports (2020). https://doi.org/10.1038/s41598-020-68622-x
59. Konopka, Joanna K.; Chatterjee, Pranab; LaMontagne, Connor; Brown, Joe. Environmental impacts of mass drug administration programs: exposures, risks, and mitigation of antimicrobial resistance. Infectious Diseases of Poverty (2022). https://doi.org/10.1186/s40249-022-01000-z
60. Rubin, Daniel H. F.; Ma, Kevin C.; Westervelt, Kathleen A.; Hullahalli, Karthik; Waldor, Matthew K.; Grad, Yonatan H. CanB is a metabolic mediator of antibiotic resistance in Neisseria gonorrhoeae. Nature Microbiology (2023). https://doi.org/10.1038/s41564-022-01282-x
61. Holt, G. S.; Lodge, J. K.; McCarthy, A. J.; Graham, A. K.; Young, G.; Bridge, S. H.; Brown, A. K.; Veses-Garcia, M.; Lanyon, C. V.; Sails, A.; Allison, H. E.; Smith, D. L. Shigatoxin encoding Bacteriophage $\phi 24_B$ modulates bacterial metabolism to raise antimicrobial tolerance. Scientific Reports (2017). https://doi.org/10.1038/srep40424
62. Pinheiro, Fernanda; Warsi, Omar; Andersson, Dan I.; Lässig, Michael. Metabolic fitness landscapes predict the evolution of antibiotic resistance. Nature Ecology & Evolution (2021). https://doi.org/10.1038/s41559-021-01397-0
63. Palomino, Alana; Gewurz, Danya; DeVine, Lela; Zajmi, Ujana; Moralez, Jenifer; Abu-Rumman, Fatima; Smith, Robert P.; Lopatkin, Allison J. Metabolic genes on conjugative plasmids are highly prevalent in Escherichia coli and can protect against antibiotic treatment. The ISME Journal: Multidisciplinary Journal of Microbial Ecology (2022). https://doi.org/10.1038/s41396-022-01329-1

Chapter 8
Immunometabolic Responses upon Bacterial Infections

Vijay Soni and Aditya Upadhyay

8.1 Introduction by the Editor

Immunometabolism—the intersection of immune signaling and metabolic regulation—has emerged as a critical field in understanding host–pathogen interactions. During bacterial infections, immune cells undergo profound metabolic reprogramming that not only fuels their defensive functions but also shapes the outcome of infection. This chapter explores the dynamic and reciprocal relationship between bacterial pathogens and host immunometabolic responses, highlighting how metabolic shifts in immune cells influence pathogen survival, inflammation, and disease progression.

Iron metabolism is a central theme in immunometabolic regulation. Iron is essential for both host and pathogen survival, and its homeostasis is tightly controlled during infection. The study by Crielaard et al. emphasizes the therapeutic potential of targeting iron metabolism, particularly through modulation of hepcidin and ferroportin pathways. These iron-regulatory mechanisms are implicated in aging-related diseases, cancer, and chronic inflammation. By influencing macrophage iron recycling, novel strategies may emerge to treat infections and inflammatory disorders where iron availability plays a pivotal role.

Macrophages are key players in immunometabolism, and their metabolic flexibility determines their functional phenotype. The study by Wei et al. reveals that the laccase domain-containing protein FAMIN (LACC1) bridges nitric oxide synthase (NOS2) activity and polyamine metabolism in inflammatory macrophages. FAMIN catalyzes the conversion of L-citrulline to L-ornithine and isocyanic acid, linking

V. Soni (✉)
Department of Medicine, Weill Cornell Medicine, New York, NY, USA
e-mail: vis2032@med.cornell.edu

A. Upadhyay
Hospital for Special Surgery, New York, NY, USA

V. Soni, A. Upadhyay (eds.), *Metabolism of Bacterial Infection*,
https://doi.org/10.1007/978-3-031-46895-7_8

arginine metabolism to polyamine synthesis. These metabolites are crucial for macrophage function and bacterial clearance. Deficiency in FAMIN leads to exaggerated inflammation and impaired immunity, suggesting its potential as a therapeutic target in diseases like Crohn's and juvenile idiopathic arthritis.

Pathogens have evolved mechanisms to exploit host immunometabolic pathways. *Mycobacterium tuberculosis* (Mtb), for instance, induces the expression of miR-33 and miR-33* to suppress autophagy and reprogram lipid metabolism in macrophages. As shown by Ouimet et al., these microRNAs inhibit key autophagy regulators and promote lipid accumulation, creating a nutrient-rich niche for Mtb survival. Inhibiting miR-33 enhances autophagic flux and reduces bacterial burden, highlighting the therapeutic promise of miRNA-based interventions in tuberculosis.

The complexity of macrophage–pathogen interactions is further explored by Russell et al., who emphasize the need for in vivo models to study immunometabolism. Using dual RNA sequencing and fluorescent fitness reporters, they demonstrate how macrophage lineage and polarization influence pathogen replication. Pathogens like *Salmonella*, *Legionella*, and *Listeria* adapt their metabolism to exploit host nutrient compartments, while macrophage glycolytic states can either support or restrict bacterial growth. These findings underscore the importance of macrophage heterogeneity in infection control and drug tolerance.

The microbiome also plays a crucial role in shaping immunometabolic responses. Zheng et al. review how early-life colonization by commensal microbes influences immune system maturation and long-term health. Factors such as delivery mode, maternal antibodies, and diet affect microbiome composition and immune development. Dysbiosis during critical developmental windows can predispose individuals to immune-mediated diseases, suggesting that microbiome-targeted therapies may restore immune balance and prevent chronic inflammation.

Staphylococcus aureus exemplifies how pathogens adapt to immunometabolic pressures. Tomlinson et al. show that itaconate, a host-derived metabolite, drives metabolic reprogramming in *S. aureus*, promoting biofilm formation and persistence in the lung. This adaptation involves decreased glycolysis and increased gluconeogenesis, favoring chronic infection. Understanding how pathogens respond to immunometabolites like itaconate may inform strategies to disrupt biofilm-associated resistance.

Finally, Rosenberg et al. provide a comprehensive overview of immunometabolic crosstalk during bacterial infection. They detail how macrophage activation leads to metabolic shifts, including accumulation of citrate and itaconate, which regulate immune responses and pathogen clearance. However, some bacteria degrade these metabolites to survive, illustrating the evolutionary arms race between host defenses and microbial adaptation.

Together, these studies reveal that immunometabolism is a dynamic and bidirectional process that shapes the outcome of bacterial infections. This chapter provides a comprehensive overview of how metabolic reprogramming in immune cells and pathogens influences inflammation, immunity, and disease progression, offering insights into novel therapeutic strategies targeting metabolic pathways.

8.2 Machine Generated Summaries

Disclaimer: The summaries in this chapter were generated from Springer Nature publications using abstractive AI auto-summarization: An abstraction-based summarizer creates new text based on deep learning. New phrases are created to summarize the content. As the constituted sentences are machine selected, they may not fully reflect the body of the work, so we strongly advise that the original content is read and cited. The auto-generated summaries were curated by the editor to meet Springer Nature publication standards.

Targeting Iron Metabolism in Drug Discovery and Delivery [125]

This is a machine-generated summary of:

Crielaard, Bart J.; Lammers, Twan; Rivella, Stefano Targeting Iron Metabolism in Drug Discovery and Delivery [125]

Published in: Nature Reviews Drug Discovery (2017)

Link to original: https://doi.org/10.1038/nrd.2016.248

If you want to cite the papers, please refer to the original.

For technical reasons we could not place the page where the original quote is coming from.

Research Highlights

Iron is a crucial element for life, playing vital roles in various biological processes, including DNA replication and oxygen transport. Its ability to transfer electrons makes it an essential cofactor in numerous biochemical reactions. However, imbalances in iron metabolism can lead to severe health issues, such as iron-deficiency anemia and iron-loading anemia. These conditions highlight the importance of maintaining proper iron homeostasis. Now, a new study by Bart J. Crielaard and colleagues explores the potential of targeting iron metabolism as a therapeutic strategy for a range of diseases beyond traditional hematological disorders.

The study provides a comprehensive overview of iron regulatory pathways and their key players, emphasizing the role of hepcidin as the master regulator of systemic iron metabolism. Hepcidin controls iron levels by modulating ferroportin, the only known cellular exporter of unbound iron in vertebrates. The authors discuss how dysregulation in these pathways is linked to various disorders, including aging-related diseases like neurodegeneration and cancer. The research also delves into cellular iron trafficking mechanisms, highlighting the importance of transferrin-mediated uptake and the role of non-transferrin-bound iron in cellular processes. By examining these pathways, the study identifies potential therapeutic targets for diseases associated with dysfunctional iron homeostasis.

The main advances from this study suggest that targeting iron metabolism could offer new therapeutic options for difficult-to-treat disorders beyond hematological conditions. Future studies could focus on developing therapies that modulate macrophage activity by interfering with their role in iron recycling, which may be particularly beneficial for cancer and chronic inflammatory disorders. Additionally, understanding the relationship between macrophage phenotype and its iron load could lead to novel strategies for macrophage-modulating therapies. Overall, this research underscores the potential of targeting iron metabolism as a promising approach for treating a wide range of diseases associated with disturbed iron homeostasis.

LACC1 Bridges NOS2 and Polyamine Metabolism in Inflammatory Macrophages [126]

This is a machine-generated summary of:

Wei, Zheng; Oh, Joonseok; Flavell, Richard A.; Crawford, Jason M. LACC1 Bridges NOS2 and Polyamine Metabolism in Inflammatory Macrophages [126]

Published in: Nature (2022)

Link to original: https://doi.org/10.1038/s41586-022-05111-3

If you want to cite the papers, please refer to the original.

For technical reasons we could not place the page where the original quote is coming from.

Research Highlights

The innate immune system serves as the body's first line of defense against infections and injuries, primarily through the activation of pattern recognition receptors that initiate inflammatory responses. Reactive nitrogen species (RNS) play a crucial role in regulating various cell types, with inducible nitric oxide synthase (iNOS) being a major source of RNS in macrophages. Despite the known roles of iNOS products like nitric oxide and L-citrulline, their complete metabolic fate remains unclear. Now, a new study by Zheng Wei and colleagues explores the biochemical function of the laccase domain-containing protein FAMIN, revealing its role as an isocyanic acid synthase in inflammatory macrophages.

The study employed a combination of in vitro biochemical assays and mouse models to elucidate FAMIN's function. Using recombinant FAMIN, the researchers identified it as a zinc-dependent metalloenzyme that cleaves L-citrulline to produce L-ornithine and isocyanic acid (HNCO). This enzymatic activity was confirmed through mass spectrometry and nuclear magnetic resonance analyses. The team further demonstrated that FAMIN's enzymatic products contribute to polyamine synthesis, which is crucial for macrophage function during bacterial infections. In mouse models deficient in FAMIN, there was an exacerbated inflammatory response and impaired bacterial clearance, which could be partially rescued by supplementing with L-ornithine.

The findings suggest that FAMIN plays a central role in macrophage immunometabolism by linking L-arginine metabolism to polyamine synthesis and HNCO production. This study provides insights into how FAMIN regulates inflammation and antibacterial responses, highlighting its potential as a target for therapeutic interventions in inflammatory diseases. Future studies could explore the broader implications of FAMIN's enzymatic activity on immune regulation and its potential as a biomarker or therapeutic target for conditions like Crohn’s disease and juvenile idiopathic arthritis.

Mycobacterium tuberculosis *Induces the miR-33 Locus to Reprogram Autophagy and Host Lipid Metabolism [127]*

This is a machine-generated summary of:

Ouimet, Mireille; Koster, Stefan; Sakowski, Erik; Ramkhelawon, Bhama; van Solingen, Coen; Oldebeken, Scott; Karunakaran, Denuja; Portal-Celhay, Cynthia; Sheedy, Frederick J; Ray, Tathagat Dutta; Cecchini, Katharine; Zamore, Philip D; Rayner, Katey J; Marcel, Yves L; Philips, Jennifer A; Moore, Kathryn J Mycobacterium Tuberculosis Induces the miR-33 Locus to Reprogram Autophagy and Host Lipid Metabolism [127]

Published in: Nature Immunology (2016)

Link to original: https://doi.org/10.1038/ni.3434

If you want to cite the papers, please refer to the original.

For technical reasons we could not place the page where the original quote is coming from.

Research Highlights

Tuberculosis, caused by *Mycobacterium tuberculosis* (Mtb), remains a leading cause of bacterial mortality worldwide. Mtb's ability to survive within macrophages by evading lysosomal degradation is a key factor in its persistence. Autophagy, a cellular process that degrades intracellular pathogens, plays a crucial role in immune responses by promoting antigen presentation and modulating inflammation. However, Mtb has developed strategies to evade autophagy, limiting its degradation. Now, a new study by Mireille Ouimet and colleagues reveals that Mtb exploits host microRNAs (miRNAs) to impair autophagy and enhance its survival within macrophages.

The study employed various experimental approaches to investigate the role of miRNAs in Mtb infection. The researchers found that Mtb infection induces the expression of miR-33 and its passenger strand miR-33*, which target key genes involved in the autophagy pathway. Using mouse and human macrophage models, they demonstrated that these miRNAs suppress autophagy by repressing genes encoding autophagy effectors and transcriptional regulators such as AMPK and TFEB. The study also showed that miR-33 and miR-33* promote lipid accumulation in macrophages by inhibiting fatty acid oxidation, providing a nutrient-rich environment for Mtb. Inhibition of these miRNAs enhanced autophagic flux and reduced Mtb survival, highlighting their role in modulating host–pathogen interactions.

The findings of this study provide significant insights into the molecular mechanisms employed by Mtb to evade host defenses. By identifying miR-33 and miR-33* as key regulators of autophagy and lipid metabolism during infection, the research highlights potential therapeutic targets for enhancing host immunity against Mtb. Future studies could explore the development of miRNA-based interventions to boost autophagic responses and improve tuberculosis treatment outcomes. Additionally, understanding the broader implications of miRNA regulation in other infectious diseases could further elucidate their role in immune modulation.

Immunometabolism at the Interface Between Macrophages and Pathogens [128]

This is a machine-generated summary of:

Russell, David G.; Huang, Lu; VanderVen, Brian C. Immunometabolism at the Interface Between Macrophages and Pathogens [128]

Published in: Nature Reviews Immunology (2019)

Link to original: https://doi.org/10.1038/s41577-019-0124-9

If you want to cite the papers, please refer to the original.

For technical reasons we could not place the page where the original quote is coming from.

Research Highlights

Macrophage metabolism is highly adaptable and reflects the pathology of various diseases, such as atherosclerosis and cancer. Traditionally, it was believed that macrophages originated from a common blood lineage and were polarized into two states based on cytokine exposure. However, recent insights reveal that macrophages arise from distinct lineages during embryonic development, leading to diverse metabolic responses to tissue injury and disease. Now, a new study by David G. Russell and colleagues explores the complex interplay between macrophage metabolism and intracellular pathogens, emphasizing the limitations of in vitro models and highlighting advanced tools like dual RNA sequencing for in vivo investigations.

The study employs cutting-edge methodologies to unravel the metabolic interactions between host macrophages and bacterial pathogens. By utilizing dual RNA sequencing at the single-cell level and fluorescent bacterial fitness reporters, the researchers investigate how different macrophage lineages respond metabolically to infections in vivo. The study focuses on bacterial pathogens such as *Salmonella Typhimurium*, *Legionella pneumophila*, *Mycobacterium tuberculosis*, and *Listeria monocytogenes*. These pathogens exhibit diverse metabolic strategies to survive within macrophages, including flexible carbon source utilization and compartmentalized nutrient use. The research highlights how macrophage polarization affects pathogen growth, with glycolytic macrophages supporting some pathogens' replication while others are restricted by nutrient limitation mechanisms like iron sequestration.

The findings underscore the importance of understanding macrophage heterogeneity in infection control. Future study directions include exploring how specific macrophage lineages influence pathogen persistence and drug tolerance. The integration of advanced tools like fluorescent fitness reporters and dual RNA-seq will enhance our understanding of host–pathogen interactions at a cellular level. This knowledge could inform the development of more effective vaccines and therapeutic strategies targeting both host metabolism and pathogen survival mechanisms. The study emphasizes the need for models that reflect in vivo complexity to accurately assess the metabolic crosstalk between host cells and pathogens.

Interaction Between Microbiota and Immunity in Health and Disease [129]

This is a machine-generated summary of:

Zheng, Danping; Liwinski, Timur; Elinav, Eran Interaction Between Microbiota and Immunity in Health and Disease [129]

Published in: Cell Research (2020)

Link to original: https://doi.org/10.1038/s41422-020-0332-7

Copyright of the summarized publication:

If you want to cite the papers, please refer to the original.

For technical reasons we could not place the page where the original quote is coming from.

Research Highlights

The human body is home to a vast array of microorganisms, collectively known as the microbiome, which inhabit various environments such as the gut and skin. Over the past two decades, advancements in genomic techniques have allowed for a deeper understanding of these microorganisms and their genomes. Recent research has highlighted the active role of the gut microbiome in influencing host functions, including metabolism and immunity. The mammalian immune system, comprising innate and adaptive components, has coevolved with these microorganisms to maintain mutualism and homeostasis. However, disruptions in this balance can lead to diseases such as inflammatory bowel disease and rheumatoid arthritis. Now, a new study by Danping Zheng and colleagues explores the intricate interactions between the microbiome and the immune system, providing insights into their roles in health and disease.

Zheng et al. conducted a comprehensive review of current knowledge on microbiome–immune interactions, focusing on both homeostatic and diseased states. The study utilized germ-free animal models to investigate mechanistic causal relationships between commensal microbiota and host immunity. Findings revealed that early-life colonization plays a crucial role in immune system maturation, with microbial diversity during this period being critical for establishing an immunoregulatory network. The study also examined how various factors such as delivery mode and maternal antibodies influence initial colonization. Additionally, it highlighted the role of specific bacteria like segmented filamentous bacteria in inducing immune responses. The research further delved into the impact of environmental factors like antibiotics and diet on microbiome composition and subsequent immune responses.

The study by Zheng et al. underscores the importance of early-life microbiome–immune interactions in shaping long-term health outcomes. However, it also points out that many mechanisms underlying these interactions remain poorly defined.

Future studies are needed to explore the long-term impacts of subtle dysbiosis during neonatal periods on adult immunity and disease risk. The research suggests that understanding these modulatory effects could have significant implications for preventing and treating immune-related disorders. Additionally, it highlights the potential for developing microbiome-targeted therapeutic strategies, such as fecal microbiota transplantation and next-generation probiotics, to restore healthy microbiome configurations in patients with immune-mediated diseases.

Staphylococcus aureus *Induces an Itaconate-Dominated Immunometabolic Response that Drives Biofilm Formation [130]*

This is a machine-generated summary of:

Tomlinson, Kira L.; Lung, Tania Wong Fok; Dach, Felix; Annavajhala, Medini K.; Gabryszewski, Stanislaw J.; Groves, Ryan A.; Drikic, Marija; Francoeur, Nancy J.; Sridhar, Shwetha H.; Smith, Melissa L.; Khanal, Sara; Britto, Clemente J.; Sebra, Robert; Lewis, Ian; Uhlemann, Anne-Catrin; Kahl, Barbara C.; Prince, Alice S.; Riquelme, Sebastián A. *Staphylococcus aureus* Induces an Itaconate-Dominated Immunometabolic Response That Drives Biofilm Formation [130]

Published in: Nature Communications (2021)

Link to original: https://doi.org/10.1038/s41467-021-21718-y

If you want to cite the papers, please refer to the original.

For technical reasons we could not place the page where the original quote is coming from.

Research Highlights

Staphylococcus aureus is a significant Gram-positive pathogen responsible for various acute and chronic infections, particularly in the lungs. The persistence of *S. aureus* in the airway, especially in conditions like chronic obstructive pulmonary disease and ventilator-associated pneumonia, is facilitated by its ability to adapt to the host's immune response. Understanding these adaptive mechanisms is crucial in the context of increasing multidrug resistance. Now, a new study by Kira L. Tomlinson and colleagues explores how *S. aureus* adapts to the inflammatory environment of the airway, focusing on the role of itaconate, a metabolite produced during immune responses.

The study employed a combination of metabolomics, transcriptomics, and genetic analysis to investigate how *S. aureus* responds to itaconate in the airway. Mice were infected with S. aureus, and bronchoalveolar lavage fluid was analyzed to assess metabolic changes. The researchers found that *S. aureus* infection specifically induced itaconate production without increasing succinate levels, unlike infections with Gram-negative bacteria like *Pseudomonas aeruginosa*. Itaconate exposure led to significant transcriptomic and metabolic changes in *S. aureus*, including decreased glycolysis and increased gluconeogenesis, promoting biofilm formation over planktonic growth. These findings were corroborated by analyzing clinical isolates from chronically infected patients, which showed similar metabolic adaptations.

The study advances our understanding of *S. aureus* pathogenesis by highlighting how itaconate acts as a selective pressure that drives metabolic reprogramming in the bacteria, favoring biofilm formation and persistence in the lung environment. Future studies could explore potential therapeutic strategies targeting these metabolic pathways to combat chronic *S. aureus* infections effectively. Additionally, further research may investigate whether itaconate induces posttranslational modifications in other bacterial proteins, contributing to its role in persistent infections. These insights into host–pathogen interactions underscore the complexity of bacterial adaptation mechanisms and their implications for treatment strategies against resistant strains of *S. aureus*.

Immunometabolic Crosstalk During Bacterial Infection [131]

This is a machine-generated summary of:

Rosenberg, Gili; Riquelme, Sebastian; Prince, Alice; Avraham, Roi Immunometabolic Crosstalk During Bacterial Infection [131]

Published in: Nature Microbiology (2022)

Link to original: https://doi.org/10.1038/s41564-022-01080-5

If you want to cite the papers, please refer to the original.

For technical reasons we could not place the page where the original quote is coming from.

Research Highlights

Interactions between pathogens and their hosts lead to metabolic changes that influence infection outcomes. Host immune cells, such as macrophages, detect pathogens through pattern-recognition receptors, triggering immune responses. Over the past decade, research has shown that immune-cell activation is linked to changes in energy metabolism, termed "immunometabolism," which regulates cytokine and chemokine expression and antimicrobial processes. Recent evidence suggests that bacterial pathogens can sense and exploit immunometabolites to enhance their survival. Now, a new study by Gili Rosenberg and colleagues explores the implications of immunometabolism on bacterial infections, focusing on how macrophage metabolism influences pathogen survival.

The study investigates the dynamic activation of macrophages, which can polarize into pro- or anti-inflammatory states upon detecting bacterial components via Toll-like receptors (TLRs). This recognition triggers signal transduction cascades that reprogram macrophage energy metabolism from oxidative phosphorylation to aerobic glycolysis. This shift leads to the accumulation of immunometabolites like citrate and itaconate, which play roles in immune activation and pathogen clearance. The researchers detail how metabolic reprogramming involves disruptions in the tricarboxylic acid (TCA) cycle, leading to increased citrate and *cis*-aconitate levels due to transcriptional repression of isocitrate dehydrogenase and nitric oxide synthase activity. Itaconate acts as an antimicrobial agent within macrophages but also poses a challenge as some bacteria have developed mechanisms to degrade itaconate for survival.

The study highlights significant advances in understanding the metabolic interplay between host immune responses and bacterial adaptation strategies. It reveals how pathogens like *Pseudomonas aeruginosa* and *Staphylococcus aureus* adapt to host-derived metabolites such as succinate and itaconate during chronic infections. These adaptations include shifts in metabolic pathways that support biofilm formation and antibiotic resistance. Future studies could focus on targeting these metabolic interactions to develop novel therapeutic strategies against persistent infections. Understanding these complex host–pathogen interactions may pave the way for innovative treatments that manipulate host or bacterial metabolism to combat infectious diseases effectively.

8.3 Conclusion

The studies presented in this chapter demonstrate the central role of immunometabolism in shaping host–pathogen interactions during bacterial infections. Far from being passive biochemical processes, metabolic pathways in immune cells are actively reprogrammed in response to microbial signals, influencing inflammation, pathogen clearance, and disease outcomes. Simultaneously, pathogens have evolved sophisticated strategies to exploit or evade these metabolic shifts, creating a dynamic interplay that determines the trajectory of infection. This chapter highlights the complexity and therapeutic potential of targeting immunometabolic pathways to modulate immune responses and combat infectious diseases.

Iron metabolism emerges as a critical axis in immunometabolic regulation. The work by Crielaard et al. illustrates how iron homeostasis, governed by hepcidin and ferroportin, influences macrophage function and systemic immunity. Dysregulation of iron pathways is implicated in chronic inflammation, neurodegeneration, and cancer. Therapeutic strategies that modulate iron recycling in macrophages may offer new avenues for treating infections and inflammatory disorders, particularly by limiting iron availability to pathogens while preserving host immune function.

The enzymatic activity of FAMIN (LACC1), as described by Wei et al., bridges nitric oxide production and polyamine synthesis in inflammatory macrophages. This metabolic link is essential for effective immune responses and bacterial clearance. FAMIN-deficient models exhibit heightened inflammation and impaired immunity, which can be partially rescued by supplementing with L-ornithine. These findings suggest that targeting FAMIN or its downstream metabolites could modulate macrophage function in diseases characterized by excessive inflammation or immune dysfunction.

Pathogens like *Mycobacterium tuberculosis* have evolved mechanisms to manipulate host immunometabolism for their benefit. Ouimet et al. demonstrate that Mtb induces miR-33 and miR-33* to suppress autophagy and promote lipid accumulation in macrophages, creating a favorable niche for bacterial survival. Inhibiting these miRNAs restores autophagic flux and reduces bacterial burden, highlighting the potential of miRNA-based therapies in tuberculosis and other chronic infections.

The heterogeneity of macrophage lineages and their metabolic responses to infection are explored by Russell et al., who emphasize the limitations of in vitro models and the need for in vivo investigations. Using dual RNA sequencing and fluorescent fitness reporters, they reveal how macrophage polarization affects pathogen replication and drug tolerance. These insights underscore the importance of understanding macrophage diversity in designing effective immunotherapies and vaccines.

The microbiome's influence on immunometabolism is another critical theme. Zheng et al. show that early-life microbial colonization shapes immune system development and long-term health. Disruptions in microbiome composition during critical windows can predispose individuals to immune-mediated diseases. Therapeutic strategies such as fecal microbiota transplantation and next-generation probiotics may restore immune balance and prevent chronic inflammation by modulating microbiome–immunity interactions.

Staphylococcus aureus adapts to immunometabolic pressures by responding to itaconate, a host-derived metabolite. Tomlinson et al. reveal that itaconate drives metabolic reprogramming in *S. aureus*, promoting biofilm formation and persistence in the lung. This adaptation involves a shift from glycolysis to gluconeogenesis, favoring chronic infection. Targeting these metabolic pathways may disrupt biofilm-associated resistance and improve treatment outcomes for persistent infections.

Rosenberg et al. provide a comprehensive overview of immunometabolic cross-talk, detailing how macrophage activation leads to metabolic shifts that regulate immune responses and pathogen survival. Accumulation of metabolites like citrate and itaconate plays dual roles—enhancing antimicrobial activity while also being exploited by pathogens. Understanding these interactions offers opportunities to develop therapies that manipulate host or bacterial metabolism to combat infections more effectively.

In conclusion, immunometabolism is a dynamic and integrative field that bridges immune signaling and metabolic regulation. The studies in this chapter reveal that metabolic reprogramming in immune cells and pathogens is central to infection biology, influencing inflammation, immunity, and disease progression. By targeting immunometabolic pathways, we can develop novel therapeutic strategies that enhance host defenses, disrupt pathogen survival mechanisms, and restore immune balance. As our understanding of immunometabolism deepens, it will pave the way for innovative treatments that address the metabolic foundations of infectious diseases.

Bibliography

1. Rivella, S. β-Thalassemias: paradigmatic diseases for scientific discoveries and development of innovative therapies. *Haematologica 100*, 418–430 (2015).
2. Fenton, H. J. H. Oxidation of tartaric acid in presence of iron. *J. Chem. Soc. Trans. 65*, 899–910 (1894).
3. Haber, F. & Weiss, J. Über die Katalyse des Hydroperoxydes. *Naturwissenschaften 20*, 948–950 (in German) (1932).
4. Donovan, A. et al. Positional cloning of zebrafish *ferroportin1* identifies a conserved vertebrate iron exporter. *Nature 403*, 776–781 (2000).
5. White, C. et al. HRG1 is essential for heme transport from the phagolysosome of macrophages during erythrophagocytosis. *Cell Metab. 17*, 261–270 (2013).
6. Li, J. Y. et al. Scara5 is a ferritin receptor mediating non-transferrin iron delivery. *Dev. Cell 16*, 35–46 (2009).
7. Leibold, E. A. & Munro, H. N. Cytoplasmic protein binds *in vitro* to a highly conserved sequence in the 5′ untranslated region of ferritin heavy- and light-subunit mRNAs. *Proc. Natl Acad. Sci. USA 85*, 2171–2175 (1988).
8. Rouault, T. A., Hentze, M. W., Caughman, S. W., Harford, J. B. & Klausner, R. D. Binding of a cytosolic protein to the iron-responsive element of human ferritin messenger RNA. *Science 241*, 1207–1210 (1988).
9. Gardenghi, S. et al. Distinct roles for hepcidin and interleukin-6 in the recovery from anemia in mice injected with heat-killed *Brucella abortus*. *Blood 123*, 1137–1145 (2014).
10. Weiss, G. & Schett, G. Anaemia in inflammatory rheumatic diseases. *Nat. Rev. Rheumatol. 9*, 205–215 (2013).
11. Sindrilaru, A. et al. An unrestrained proinflammatory M1 macrophage population induced by iron impairs wound healing in humans and mice. *J. Clin. Invest. 121*, 985–997 (2011). *This study is the first to directly link the inflammatory phenotype of macrophages to their iron load.*
12. Dekker, M. C. J. et al. Mutations in the hemochromatosis gene (*HFE*), Parkinson's disease and parkinsonism. *Neurosci. Lett. 348*, 117–119 (2003).

13. Wang, X.-S. et al. Increased incidence of the *Hfe* mutation in amyotrophic lateral sclerosis and related cellular consequences. *J. Neurol. Sci. 227*, 27–33 (2004).
14. Biswas, S. K., Chittezhath, M., Shalova, I. N. & Lim, J.-Y. Macrophage polarization and plasticity in health and disease. *Immunol. Res. 53*, 11–24 (2012).
15. Nairz, M. et al. 'Ride on the ferrous wheel' — the cycle of iron in macrophages in health and disease. *Immunobiology 220*, 280–294 (2015).
16. Zanganeh, S. et al. Iron oxide nanoparticles inhibit tumour growth by inducing pro-inflammatory macrophage polarization in tumour tissues. *Nat. Nanotechnol. 11*, 986–994 (2016). *This study demonstrates the capacity of iron to modulate the activity of intratumoural macrophages to an antitumoural phenotype, which could be exploited as an anticancer strategy.*
17. Bronte, V. & Murray, P. J. Understanding local macrophage phenotypes in disease: modulating macrophage function to treat cancer. *Nat. Med. 21*, 117–119 (2015).
18. Crielaard, B. J. et al. Macrophages and liposomes in inflammatory disease: friends or foes? *Int. J. Pharm. 416*, 499–506 (2011).
19. Heng, T. S. & Painter, M. W., Consortium, I. G. P. The Immunological Genome Project: networks of gene expression in immune cells. *Nat. Immunol. 9*, 1091–1094 (2008).
20. Skon-Hegg, C. et al. LACC1 regulates TNF and IL-17 in mouse models of arthritis and inflammation. *J Immunol. 202*, 183–193 (2019).
21. Kang, J. W. et al. Myeloid cell expression of LACC1 is required for bacterial clearance and control of intestinal inflammation. *Gastroenterology 159*, 1051–1067 (2020).
22. Mogensen, T. H. Pathogen recognition and inflammatory signaling in innate immune defenses. *Clin. Microbiol. Rev. 22*, 240–273 (2009).
23. Marletta, M. A., Yoon, P. S., Iyengar, R., Leaf, C. D. & Wishnok, J. S. Macrophage oxidation of L-arginine to nitrite and nitrate: nitric oxide is an intermediate. *Biochemistry 27*, 8706–8711 (1988).
24. Nathan, C. & Shiloh, M. U. Reactive oxygen and nitrogen intermediates in the relationship between mammalian hosts and microbial pathogens. *Proc. Natl Acad. Sci. USA 97*, 8841–8848 (2000).
25. Omarjee, O. et al. LACC1 deficiency links juvenile arthritis with autophagy and metabolism in macrophages. *J. Exp. Med. 218*, e20201006 (2021).
26. Murray, P. J. et al. Macrophage activation and polarization: nomenclature and experimental guidelines. *Immunity 41*, 14–20 (2014).
27. Pearce, E. L. & Pearce, E. J. Metabolic pathways in immune cell activation and quiescence. *Immunity 38*, 633–643 (2013).
28. Goedeke, L. et al. A regulatory role for microRNA 33* in controlling lipid metabolism gene expression. *Mol. Cell. Biol. 33*, 2339–2352 (2013).
29. Singh, R. & Cuervo, A.M. Autophagy in the cellular energetic balance. *Cell Metab. 13*, 495–504 (2011).
30. Ouimet, M. et al. Autophagy regulates cholesterol efflux from macrophage foam cells via lysosomal acid lipase. *Cell Metab. 13*, 655–667 (2011).
31. O'Neill, L. A. & Pearce, E. J. Immunometabolism governs dendritic cell and macrophage function. *J. Exp. Med. 213*, 15–23 (2016).
32. Van den Bossche, J., O'Neill, L. A. & Menon, D. Macrophage immunometabolism: where are we (going)? *Trends. Immunol. 38*, 395–406 (2017).
33. VanderVen, B. C., Yates, R. M. & Russell, D. G. Intraphagosomal measurement of the magnitude and duration of the oxidative burst. *Traffic 10*, 372–378 (2009).
34. O'Neill, L. A., Kishton, R. J. & Rathmell, J. A guide to immunometabolism for immunologists. *Nat. Rev. Immunol. 16*, 553–565 (2016).
35. Mosser, D. M. & Edwards, J. P. Exploring the full spectrum of macrophage activation. *Nat. Rev. Immunol. 8*, 958–969 (2008).
36. Olive, A. J. & Sassetti, C. M. Metabolic crosstalk between host and pathogen: sensing, adapting and competing. *Nat. Rev. Microbiol. 14*, 221–234 (2016).
37. Isaac, D. T. & Isberg, R. Master manipulators: an update on *Legionella pneumophila* Icm/Dot translocated substrates and their host targets. *Future Microbiol. 9*, 343–359 (2014).

38. Botella, H. et al. Mycobacterial p(1)-type ATPases mediate resistance to zinc poisoning in human macrophages. *Cell Host Microbe 10*, 248–259 (2011).
39. Crouch, M. L., Castor, M., Karlinsey, J. E., Kalhorn, T. & Fang, F. C. Biosynthesis and IroC-dependent export of the siderophore salmochelin are essential for virulence of *Salmonella enterica* serovar Typhimurium. *Mol. Microbiol. 67*, 971–983 (2008).
40. Liu, J. Z. et al. Zinc sequestration by the neutrophil protein calprotectin enhances *Salmonella* growth in the inflamed gut. *Cell Host Microbe 11*, 227–239 (2012).
41. Wolschendorf, F. et al. Copper resistance is essential for virulence of *Mycobacterium tuberculosis*. *Proc. Natl Acad. Sci. USA 108*, 1621–1626 (2011).
42. Schmidt, S. V. & Schultze, J. L. New insights into IDO biology in bacterial and viral infections. *Front. Immunol. 5*, 384 (2014).
43. Muñoz-Elías, E. J. & McKinney, J. D. *Mycobacterium tuberculosis* isocitrate lyases 1 and 2 are jointly required for *in vivo* growth and virulence. *Nature Med. 11*, 638–644 (2005).
44. Savvi, S. et al. Functional characterization of a vitamin B_{12}-dependent methylmalonyl pathway in *Mycobacterium tuberculosis*: implications for propionate metabolism during growth on fatty acids. *J. Bacteriol. 190*, 3886–3895 (2008).
45. Cheng, S. C. et al. mTOR- and HIF-1α-mediated aerobic glycolysis as metabolic basis for trained immunity. *Science 345*, 1250684 (2014).
46. Arts, R. J. W. et al. Immunometabolic pathways in BCG-induced trained immunity. *Cell Rep. 17*, 2562–2571 (2016).
47. Stapels, D. A. C. et al. *Salmonella* persisters undermine host immune defenses during antibiotic treatment. *Science 362*, 1156–1160 (2018).
48. Liu, Y. et al. Immune activation of the host cell induces drug tolerance in *Mycobacterium tuberculosis* both in vitro and in vivo. *J. Exp. Med. 213*, 809–825 (2016).
49. Gibbings, S. L. et al. Transcriptome analysis highlights the conserved difference between embryonic and postnatal-derived alveolar macrophages. *Blood 126*, 1357–1366 (2015).
50. Gibbings, S. L. et al. Three unique interstitial macrophages in the murine lung at steady state. *Am. J. Respir. Cell. Mol. Biol. 57*, 66–76 (2017).
51. Jenkins, S. J. et al. Local macrophage proliferation, rather than recruitment from the blood, is a signature of TH2 inflammation. *Science 332*, 1284–1288 (2011).
52. Jenkins, S. J. et al. IL-4 directly signals tissue-resident macrophages to proliferate beyond homeostatic levels controlled by CSF-1. *J. Exp. Med. 210*, 2477–2491 (2013).
53. Ruckerl, D. et al. Macrophage origin limits functional plasticity in helminth-bacterial co-infection. *PLOS Pathog. 13*, e1006233 (2017).
54. Macpherson, A. J., Geuking, M. B. & McCoy, K. D. Immune responses that adapt the intestinal mucosa to commensal intestinal bacteria. *Immunology 115*, 153–162 (2005).
55. Chu, H. & Mazmanian, S. K. Innate immune recognition of the microbiota promotes host-microbial symbiosis. *Nat. Immunol. 14*, 668–675 (2013).
56. Zhang, M. et al. Interactions between intestinal microbiota and host immune response in inflammatory bowel disease. *Front. Immunol. 8*, 942 (2017).
57. Valitutti, F., Cucchiara, S. & Fasano, A. Celiac disease and the microbiome. *Nutrients 11*, 2403 (2019).
58. Maeda, Y. & Takeda, K. Host-microbiota interactions in rheumatoid arthritis. *Exp. Mol. Med. 51*, 150 (2019).
59. Belizario, J. E., Faintuch, J. & Garay-Malpartida, M. Gut microbiome dysbiosis and immunometabolism: New frontiers for treatment of metabolic diseases. *Mediators Inflamm. 2018*, 2037838 (2018).
60. Main, B. S. & Minter, M. R. Microbial immuno-communication in neurodegenerative diseases. *Front. Neurosci. 11*, 151 (2017).
61. Maynard, C. L., Elson, C. O., Hatton, R. D. & Weaver, C. T. Reciprocal interactions of the intestinal microbiota and immune system. *Nature 489*, 231–241 (2012).
62. Belkaid, Y. & Harrison, O. J. Homeostatic immunity and the microbiota. *Immunity 46*, 562–576 (2017).
63. Belkaid, Y. & Hand, T. W. Role of the microbiota in immunity and inflammation. *Cell 157*, 121–141 (2014).

64. Caballero-Flores, G. et al. Maternal immunization confers protection to the offspring against an attaching and effacing pathogen through delivery of IgG in breast milk. *Cell Host Microbe 25*, 313–323 (2019).
65. Zheng, W. et al. Microbiota-targeted maternal antibodies protect neonates from enteric infection. *Nature 577*, 543–548 (2020).
66. Fulde, M. et al. Neonatal selection by Toll-like receptor 5 influences long-term gut microbiota composition. *Nature 560*, 489–493 (2018).
67. Erturk-Hasdemir, D. et al. Symbionts exploit complex signaling to educate the immune system. *Proc. Natl. Acad. Sci. USA* https://doi.org/10.1073/pnas.1915978116 (2019).
68. Vaishnava, S. et al. The antibacterial lectin RegIIIgamma promotes the spatial segregation of microbiota and host in the intestine. *Science 334*, 255–258 (2011).
69. Palm, N. W. et al. Immunoglobulin A coating identifies colitogenic bacteria in inflammatory bowel disease. *Cell 158*, 1000–1010 (2014).
70. Shulzhenko, N. et al. Crosstalk between B lymphocytes, microbiota and the intestinal epithelium governs immunity versus metabolism in the gut. *Nat. Med. 17*, 1585–1593 (2011).
71. Arpaia, N. et al. Metabolites produced by commensal bacteria promote peripheral regulatory T-cell generation. *Nature 504*, 451–455 (2013).
72. Atarashi, K. et al. Induction of colonic regulatory T cells by indigenous clostridium species. *Science 331*, 334–337 (2011).
73. Smith, P. M. et al. The microbial metabolites, short-chain fatty acids, regulate colonic Treg cell homeostasis. *Science 341*, 569–573 (2013).
74. Rescigno, M., Rotta, G., Valzasina, B. & Ricciardi-Castagnoli, P. Dendritic cells shuttle microbes across gut epithelial monolayers. *Immunobiology 204*, 572–581 (2001).
75. Russell, S. L. et al. Early life antibiotic-driven changes in microbiota enhance susceptibility to allergic asthma. *EMBO Rep. 13*, 440–447 (2012).
76. Yamamoto-Hanada, K., Yang, L., Narita, M., Saito, H. & Ohya, Y. Influence of antibiotic use in early childhood on asthma and allergic diseases at age 5. *Ann. Allergy Asthma Immunol. 119*, 54–58 (2017).
77. Kim, M. et al. Critical Role for the microbiota in CX3CR1(+) intestinal mononuclear phagocyte regulation of intestinal T cell responses. *Immunity 49*, 151–163 (2018).
78. Ohnmacht, C. et al. MUCOSAL IMMUNOLOGY. The microbiota regulates type 2 immunity through RORgammat(+) T cells. *Science 349*, 989–993 (2015).
79. Haghikia, A. et al. Dietary fatty acids directly impact central nervous system autoimmunity via the small intestine. *Immunity 43*, 817–829 (2015).
80. He, B. et al. Resetting microbiota by Lactobacillus reuteri inhibits T reg deficiency-induced autoimmunity via adenosine A2A receptors. *J. Exp. Med. 214*, 107–123 (2017).
81. Rodriguez-Palacios, A. et al. The artificial sweetener splenda promotes gut Proteobacteria, dysbiosis, and myeloperoxidase reactivity in Crohn's disease-like ileitis. *Inflamm. Bowel Dis. 24*, 1005–1020 (2018).
82. Viennois, E., Merlin, D., Gewirtz, A. T. & Chassaing, B. Dietary emulsifier-induced low-grade inflammation promotes colon carcinogenesis. *Cancer Res. 77*, 27–40 (2017).
83. Rosshart, S. P. et al. Laboratory mice born to wild mice have natural microbiota and model human immune responses. *Science 365*, eaaw4361 (2019).
84. Seregin, S. S. et al. NLRP6 protects Il10(−/−) mice from colitis by limiting colonization of Akkermansia muciniphila. *Cell Rep. 19*, 733–745 (2017).
85. Schaubeck, M. et al. Dysbiotic gut microbiota causes transmissible Crohn's disease-like ileitis independent of failure in antimicrobial defence. *Gut 65*, 225–237 (2016).
86. Nakamura, Y. et al. Staphylococcus delta-toxin induces allergic skin disease by activating mast cells. *Nature 503*, 397–401 (2013).
87. Jin, C. et al. Commensal microbiota promote lung cancer development via gammadelta T cells. *Cell 176*, 998–1013 (2019).
88. Paik, Y. H. et al. Toll-like receptor 4 mediates inflammatory signaling by bacterial lipopolysaccharide in human hepatic stellate cells. *Hepatology 37*, 1043–1055 (2003).

89. Ma, C. et al. Gut microbiome-mediated bile acid metabolism regulates liver cancer via NKT cells. *Science 360*, eaan5931 (2018).
90. Yoshimoto, S. et al. Obesity-induced gut microbial metabolite promotes liver cancer through senescence secretome. *Nature 499*, 97–101 (2013).
91. Loo, T. M. et al. Gut microbiota promotes obesity-associated liver cancer through PGE2-mediated suppression of antitumor immunity. *Cancer Discov. 7*, 522–538 (2017).
92. Limon, J. J. et al. Malassezia is associated with Crohn's disease and exacerbates colitis in mouse models. *Cell Host Microbe 25*, 377–388 (2019).
93. Sokol, H. et al. Fungal microbiota dysbiosis in IBD. *Gut 66*, 1039–1048 (2017).
94. Norman, J. M. et al. Disease-specific alterations in the enteric virome in inflammatory bowel disease. *Cell 160*, 447–460 (2015).
95. Zuo, T. et al. Gut mucosal virome alterations in ulcerative colitis. *Gut 68*, 1169–1179 (2019).
96. Levy, M., Thaiss, C. A. & Elinav, E. Metabolites: messengers between the microbiota and the immune system. *Genes Dev. 30*, 1589–1597 (2016).
97. Chen, H. et al. A forward chemical genetic screen reveals gut microbiota metabolites that modulate host physiology. *Cell 177*, 1217–1231 (2019).
98. Mills, E. & O'Neill, L. A. J. Succinate: a metabolic signal in inflammation. *Trends Cell Biol. 24*, 313–320 (2014).
99. O'Neill, L. A. J. & Artyomov, M. N. Itaconate: the poster child of metabolic reprogramming in macrophage function. *Nat. Rev. Immunol. 19*, 273–281 (2019).
100. Riquelme, S. A. et al. *Pseudomonas aeruginosa* utilizes host-derived itaconate to redirect its metabolism to promote biofilm formation. *Cell Metab.* https://doi.org/10.1016/j.cmet.2020.04.017 (2020).
101. Lemire J, Alhasawi A, Appanna VP, Tharmalingam S, Appanna VD. Metabolic defence against oxidative stress: the road less travelled so far. J Appl Microbiol. 2017;123(4):798–809. https://doi.org/10.1111/jam.13509.
102. Kulkarni, R. et al. Cigarette smoke increases *Staphylococcus aureus* biofilm formation via oxidative stress. *Infect. Immun. 80*, 3804–3811 (2012).
103. Yin, S. et al. Burn serum increases *Staphylococcus aureus* biofilm formation via oxidative stress. *Front. Microbiol. 8*, https://doi.org/10.3389/fmicb.2017.01191 (2017).
104. Gabryszewski, S. J. et al. Metabolic adaptation in methicillin-resistant *Staphylococcus aureus* pneumonia. *Am. J. Respir. Cell Mol. Biol. 61*, 185–197 (2019).
105. Vitko, N. P., Grosser, M. R., Khatri, D., Lance, T. R. & Richardson, A. R. Expanded glucose import capability. *Affords Staphylococcus aureus Optimized Glycolytic Flux Infect. 7*, e00296-00216 (2016).
106. Balasubramanian, D., Harper, L., Shopsin, B. & Torres, V. J. Staphylococcus aureus pathogenesis in diverse host environments. *Pathogens and Disease*, **ftx005**, https://doi.org/10.1093/femspd/ftx005 (2017).
107. Kawasaki, T. & Kawai, T. Toll-like receptor signaling pathways. *Front. Immunol. 5*, 461 (2014).
108. Ryan, D. G. & O'Neill, L. A. J. Krebs cycle reborn in macrophage immunometabolism. *Annu. Rev. Immunol. 38*, 289–313 (2020).
109. Tannahill, G. et al. Succinate is a danger signal that induces IL-1β via HIF-1α. *Nature 496*, 238–242 (2013).
110. Jarc, E. & Petan, T. A twist of FATe: lipid droplets and inflammatory lipid mediators. *Biochimie 169*, 69–87 (2020).
111. Liu, P.-S. et al. α-Ketoglutarate orchestrates macrophage activation through metabolic and epigenetic reprogramming. *Nat. Immunol. 18*, 985–994 (2017).
112. Tomlinson, K. L. et al. *Staphylococcus aureus* induces an itaconate-dominated immunometabolic response that drives biofilm formation. *Nat. Commun. 12*, 1399 (2021).
113. Potter, A. D. et al. Host nutrient milieu drives an essential role for aspartate biosynthesis during invasive *Staphylococcus aureus* infection. *Proc. Natl Acad. Sci USA 117*, 12394–12401 (2020).

114. Daniel, J., Maamar, H., Deb, C., Sirakova, T. D. & Kolattukudy, P. E. *Mycobacterium tuberculosis* uses host triacylglycerol to accumulate lipid droplets and acquires a dormancy-like phenotype in lipid-loaded macrophages. *PLoS Pathog. 7*, e1002093 (2011).
115. Del Portillo, P. et al. Hypoxia is not a main stress when *Mycobacterium tuberculosis* is in a dormancy-like long-chain fatty acid environment. *Front. Cell. Infect. Microbiol. 8*, 449 (2019).
116. Hackett, E. E. et al. *Mycobacterium tuberculosis* limits host glycolysis and IL-1β by restriction of PFK-M via microRNA-21. *Cell Rep. 30*, 124–136 (2020).
117. Bhaskar, A. et al. Host sirtuin 2 as an immunotherapeutic target against tuberculosis. *eLife 9*, e55415 (2020).
118. Jiang, L. et al. *Salmonella* Typhimurium reprograms macrophage metabolism via T3SS effector SopE2 to promote intracellular replication and virulence. *Nat. Commun. 12*, 879 (2021).
119. Eisele, N. A. et al. *Salmonella* require the fatty acid regulator PPARδ for the establishment of a metabolic environment essential for long term persistence. *Cell Host Microbe 14*, 171–182 (2013).
120. Pisu, D., Huang, L., Grenier, J. K. & Russell, D. G. Dual RNA-Seq of Mtb-infected macrophages in vivo reveals ontologically distinct host–pathogen interactions. *Cell Rep. 30*, 335–350 (2020).
121. Galván-Peña, S. & O'Neill, L. A. J. Metabolic reprograming in macrophage polarization. *Front. Immunol. 5*, 420 (2014).
122. Lachmandas, E. et al. Microbial stimulation of different Toll-like receptor signalling pathways induces diverse metabolic programmes in human monocytes. *Nat. Microbiol. 2*, 16246 (2016).
123. Crielaard, Bart J.; Lammers, Twan; Rivella, Stefano. Targeting iron metabolism in drug discovery and delivery. Nature Reviews Drug Discovery (2017). https://doi.org/10.1038/nrd.2016.248
124. Wei, Zheng; Oh, Joonseok; Flavell, Richard A.; Crawford, Jason M. LACC1 bridges NOS2 and polyamine metabolism in inflammatory macrophages. Nature (2022). https://doi.org/10.1038/s41586-022-05111-3
125. Ouimet, Mireille; Koster, Stefan; Sakowski, Erik; Ramkhelawon, Bhama; van Solingen, Coen; Oldebeken, Scott; Karunakaran, Denuja; Portal-Celhay, Cynthia; Sheedy, Frederick J; Ray, Tathagat Dutta; Cecchini, Katharine; Zamore, Philip D; Rayner, Katey J; Marcel, Yves L; Philips, Jennifer A; Moore, Kathryn J. Mycobacterium tuberculosis induces the miR-33 locus to reprogram autophagy and host lipid metabolism. Nature Immunology (2016). https://doi.org/10.1038/ni.3434
126. Russell, David G.; Huang, Lu; VanderVen, Brian C. Immunometabolism at the interface between macrophages and pathogens. Nature Reviews Immunology (2019). https://doi.org/10.1038/s41577-019-0124-9
127. Zheng, Danping; Liwinski, Timur; Elinav, Eran. Interaction between microbiota and immunity in health and disease. Cell Research (2020). https://doi.org/10.1038/s41422-020-0332-7
128. Tomlinson, Kira L.; Lung, Tania Wong Fok; Dach, Felix; Annavajhala, Medini K.; Gabryszewski, Stanislaw J.; Groves, Ryan A.; Drikic, Marija; Francoeur, Nancy J.; Sridhar, Shwetha H.; Smith, Melissa L.; Khanal, Sara; Britto, Clemente J.; Sebra, Robert; Lewis, Ian; Uhlemann, Anne-Catrin; Kahl, Barbara C.; Prince, Alice S.; Riquelme, Sebastián A. Staphylococcus aureus induces an itaconate-dominated immunometabolic response that drives biofilm formation. Nature Communications (2021). https://doi.org/10.1038/s41467-021-21718-y
129. Rosenberg, Gili; Riquelme, Sebastian; Prince, Alice; Avraham, Roi. Immunometabolic crosstalk during bacterial infection. Nature Microbiology (2022). https://doi.org/10.1038/s41564-022-01080-5

Chapter 9
Metabolic Modeling (Computational and Mathematical) of Bacterial Infections

Vijay Soni and Aditya Upadhyay

9.1 Introduction by the Editor

Metabolic modeling has become an indispensable tool in modern microbiology, offering a systems-level perspective on bacterial physiology, pathogenesis, and therapeutic targeting. By integrating computational and mathematical frameworks with experimental data, researchers can simulate and predict metabolic behaviors under various conditions, including infection, drug treatment, and environmental stress. This chapter explores the diverse applications of metabolic modeling in the study of bacterial infections, highlighting how genome-scale models, machine learning, and systems biology approaches are transforming our understanding of microbial metabolism and its role in disease.

One of the most significant advancements in metabolomics is the development of high-throughput workflows for metabolite annotation. The study by Bilbao et al. introduces PeakDecoder, a machine learning-based algorithm integrated with multidimensional LC-IM-MS (liquid chromatography–ion mobility–mass spectrometry). This approach enables confident identification of metabolites in engineered microbial strains, such as *Pseudomonas putida* and *Aspergillus niger*, without relying on spectral libraries. By generating decoys and applying targeted data extraction, PeakDecoder enhances metabolite coverage and annotation accuracy, accelerating the design-build-test-learn cycle in synthetic biology and biomanufacturing.

Genome-scale metabolic models (GSMMs) are central to computational microbiology. Chung et al. review the use of GSMMs to identify novel antimicrobial

V. Soni (✉)
Department of Medicine, Weill Cornell Medicine, New York, NY, USA
e-mail: vis2032@med.cornell.edu

A. Upadhyay
Hospital for Special Surgery, New York, NY, USA

V. Soni, A. Upadhyay (eds.), *Metabolism of Bacterial Infection*,
https://doi.org/10.1007/978-3-031-46895-7_9

targets in multidrug-resistant Gram-negative pathogens. These models integrate multi-omics data to simulate bacterial physiology and predict drug responses. The reconstruction process involves automated tools like Model SEED and CarveMe, followed by manual curation to ensure biological accuracy. GSMMs provide a platform for exploring adaptive resistance mechanisms and designing synergistic drug combinations, offering a promising strategy to combat antimicrobial resistance (AMR).

The utility of GSMMs is further demonstrated in the study by Jenior et al., which investigates metabolic differences between clinical and laboratory isolates of *Klebsiella pneumoniae*. By integrating transcriptomic data with metabolic network reconstructions, the researchers identify increased reliance on valine catabolism in clinical isolates. This metabolic adaptation may contribute to bacterial survival and virulence during infection, suggesting valine metabolism as a potential therapeutic target. The study exemplifies how systems biology can uncover context-specific metabolic traits relevant to pathogenesis.

Beyond human pathogens, metabolic modeling is also applied in aquaculture. Devika et al. use GSMMs to identify microbial biomarkers in *Penaeus vannamei*, a commercially important shrimp species. By combining amplicon sequencing with constraint-based modeling, they predict beneficial microbial interactions that could limit pathogen growth and improve shrimp health. This integrative approach supports the development of probiotic formulations and sustainable aquaculture practices.

Metabolic modeling also sheds light on host–microbe interactions. Mirhakkak et al. construct a GSMM for *Candida albicans* to explore its metabolic interplay with gut bacteria. Their findings reveal that specific bacterial species can modulate fungal growth through metabolite exchange, influencing colonization levels. This research opens avenues for microbiome-targeted therapies to prevent opportunistic fungal infections.

Dynamic fluctuations in bacterial metabolism are another area of interest. Bi et al. investigate pyruvate oscillations in *Escherichia coli* using FRET-based sensors and kinetic modeling. Their study reveals that metabolic networks exhibit stochastic behavior influenced by allosteric regulation and energy state. Understanding these fluctuations may provide insights into bacterial adaptability and metabolic control mechanisms.

In the context of tuberculosis, Fang et al. present a systems biology framework that integrates enzyme inhibition kinetics with metabolic modeling to simulate drug effects on *Mycobacterium tuberculosis*. Their model accurately predicts dose–response curves for metabolic inhibitors, offering a quantitative tool for evaluating therapeutic strategies and identifying drug targets.

Finally, Sutter et al. expand the catalog of bacterial microcompartments (BMCs), protein-based organelles that encapsulate metabolic enzymes. By analyzing genomic datasets, they identify over 6000 BMC loci, doubling the known diversity and distribution. These findings highlight the modular nature of BMCs and their potential applications in metabolic engineering and synthetic biology.

Together, these studies demonstrate the power of metabolic modeling to unravel the complexities of bacterial metabolism, host–pathogen interactions, and therapeutic design. This chapter provides a comprehensive overview of computational and mathematical approaches that are reshaping our understanding of bacterial infections and guiding the development of innovative solutions in medicine and biotechnology.

9.2 Machine Generated Summaries

Disclaimer: The summaries in this chapter were generated from Springer Nature publications using abstractive AI auto-summarization: An abstraction-based summarizer creates new text based on deep learning. New phrases are created to summarize the content. As the constituted sentences are machine selected, they may not fully reflect the body of the work, so we strongly advise that the original content is read and cited. The auto-generated summaries were curated by the editor to meet Springer Nature publication standards.

PeakDecoder Enables Machine Learning-Based Metabolite Annotation and Accurate Profiling in Multidimensional Mass Spectrometry Measurements [108]

This is a machine-generated summary of:

Bilbao, Aivett; Munoz, Nathalie; Kim, Joonhoon; Orton, Daniel J.; Gao, Yuqian; Poorey, Kunal; Pomraning, Kyle R.; Weitz, Karl; Burnet, Meagan; Nicora, Carrie D.; Wilton, Rosemarie; Deng, Shuang; Dai, Ziyu; Oksen, Ethan; Gee, Aaron; Fasani, Rick A.; Tsalenko, Anya; Tanjore, Deepti; Gardner, James; Smith, Richard D.; Michener, Joshua K.; Gladden, John M.; Baker, Erin S.; Petzold, Christopher J.; Kim, Young-Mo; Apffel, Alex; Magnuson, Jon K.; Burnum-Johnson, Kristin E. PeakDecoder Enables Machine Learning-Based Metabolite Annotation and Accurate Profiling in Multidimensional Mass Spectrometry Measurements [108]

Published in: Nature Communications (2023)

Link to original: https://doi.org/10.1038/s41467-023-37031-9

If you want to cite the papers, please refer to the original.

For technical reasons we could not place the page where the original quote is coming from.

Research Highlights

Metabolomics, the study of small molecules produced by cellular processes, provides insights into physiological conditions and environmental effects on organisms. Advances in synthetic biology and genome editing have enabled the design of novel strains for biomanufacturing, necessitating accurate analytical tools for developing microorganisms that produce desired chemicals from renewable feedstocks. However, the structural diversity of metabolites poses significant challenges in detection and annotation. Now, a new study by Aivett Bilbao and colleagues introduces a high-throughput workflow combining liquid chromatography-ion mobility-mass spectrometry (LC-IM-MS) with PeakDecoder, an algorithm for metabolite identification with error rate calculations.

The study employs a multidimensional LC-IM-MS approach to analyze microbial samples from engineered strains of *Pseudomonas putida*, *Aspergillus pseudoterreus*, *Aspergillus niger*, and *Rhodosporidium toruloides*. The researchers optimized the LC method using Automated Chromatographic Method Selection Software to maximize metabolite detection. The PeakDecoder algorithm integrates untargeted feature detection and targeted data extraction to identify metabolites without relying on spectral annotations or libraries. By generating decoys from raw DIA spectra and employing machine learning techniques, PeakDecoder enhances confidence in metabolite identification. The workflow was applied to study metabolic changes in engineered strains under various conditions, revealing insights into metabolic pathways and responses.

The integration of LC-IM-MS with PeakDecoder represents a significant advancement in metabolomics, offering increased metabolite coverage and confident annotation. This approach enables faster analysis compared to traditional methods and provides a comprehensive digital record of detectable ions for future reprocessing. Future studies could explore additional decoy generation methods or machine learning models to further enhance performance. The workflow's applicability extends beyond the Agile BioFoundry consortium, offering potential benefits for environmental and biological research by facilitating rapid testing of engineered strains and accelerating the design-build-test-learn cycle in synthetic biology applications.

Novel Antimicrobial Development Using Genome-Scale Metabolic Model of Gram-Negative Pathogens: A Review [109]

This is a machine-generated summary of:

Chung, Wan Yean; Zhu, Yan; Mahamad Maifiah, Mohd Hafidz; Shivashekaregowda, Naveen Kumar Hawala; Wong, Eng Hwa; Abdul Rahim, Nusaibah Novel Antimicrobial Development Using Genome-Scale Metabolic Model of Gram-Negative Pathogens: A Review [109]

Published in: The Journal of Antibiotics (2020)

Link to original: https://doi.org/10.1038/s41429-020-00366-2

If you want to cite the papers, please refer to the original.

For technical reasons we could not place the page where the original quote is coming from.

Research Highlights

Antimicrobial resistance (AMR) poses a significant global health threat, exacerbated by the overuse of existing antimicrobials and the emergence of multidrug-resistant (MDR) bacteria. Hospital-acquired infections, particularly those caused by Gram-negative bacteria, contribute to increased mortality and healthcare costs due to limited treatment options. The World Health Organization has prioritized the development of novel antimicrobial treatments for critical pathogens like carbapenem-resistant *Pseudomonas aeruginosa*, *Enterobacteriaceae*, and *Acinetobacter baumannii*. Now, a new study by Wan Yean Chung and colleagues explores the use of genome-scale metabolic models (GSMMs) to identify novel antimicrobial drug targets and understand adaptive resistance mechanisms in MDR bacteria.

The study employs GSMMs as a systems platform for integrating multi-omics data to predict drug targets in MDR bacteria. The authors detail a five-stage process for metabolic network reconstruction, starting with draft model creation using tools like Pathway Tools and Model SEED. The draft is refined through literature curation and verification against databases such as KEGG. The model is then converted into a computable format with defined system boundaries, followed by quality control tests like gap analysis. Finally, the reconstruction is disseminated in formats like SBML for community use. Automated tools such as Model SEED and CarveMe are highlighted for their efficiency in GSMM reconstruction, while manual curation remains essential to ensure accuracy.

The study advances our understanding of bacterial metabolism and its role in antibiotic resistance. By integrating multi-omics data, GSMMs can simulate interactions between biological components and predict physiological behaviors under antibiotic treatment. Future studies could focus on developing GSMMs that incorporate both transcriptomic and metabolomic data to bridge existing gaps between these fields. This approach may lead to new drug development paradigms by revealing novel mechanisms of antibiotic adjuvants and enabling synergistic therapeutic combinations. Overall, GSMMs offer valuable insights into host–pathogen interactions and bacterial metabolism, paving the way for innovative strategies to combat AMR.

Genome-Scale Metabolic Modeling Reveals Increased Reliance on Valine Catabolism in Clinical Isolates of Klebsiella pneumoniae *[110]*

This is a machine-generated summary of:

Jenior, Matthew L.; Dickenson, Mary E.; Papin, Jason A. Genome-Scale Metabolic Modeling Reveals Increased Reliance on Valine Catabolism in Clinical Isolates of *Klebsiella pneumoniae* [110]

Published in: npj Systems Biology and Applications (2022)

Link to original: https://doi.org/10.1038/s41540-022-00252-7

If you want to cite the papers, please refer to the original.

For technical reasons we could not place the page where the original quote is coming from.

Research Highlights

Carbapenem-resistant *Enterobacteriaceae* (CRE) have become a critical concern in healthcare due to their resistance to carbapenem antibiotics, which are often the last line of defense against severe bacterial infections. *Klebsiella pneumoniae*, a Gram-negative bacterium, is a significant contributor to CRE-related infections, with increasing resistance observed in regions like the Eastern Mediterranean and Europe. The rapid acquisition of antibiotic resistance by *K. pneumoniae* necessitates the development of new treatment strategies. Now, a new study by Matthew

L. Jenior and colleagues explores potential therapeutic targets by examining metabolic differences between clinical and laboratory isolates of *K. pneumoniae*.

The study employed a meta-analysis of publicly available RNA-Seq datasets to compare transcriptional profiles of *K. pneumoniae* isolates from clinical and laboratory settings. By integrating these transcriptomic data with genome-scale metabolic network reconstructions (GENREs), the researchers identified distinct metabolic pathways active in clinical isolates. They focused on reactions unique to clinical isolates, revealing increased transcription of genes involved in aminoglycoside degradation and histidine utilization. The integration with GENREs allowed for the identification of conserved metabolic components specific to infection contexts, highlighting elevated uptake and utilization of L-valine in clinical isolates through an *Enterobacteriaceae*-specific valine transaminase.

The findings suggest that valine catabolism plays a crucial role in the metabolism of *K. pneumoniae* during infection, potentially offering new avenues for therapeutic intervention. The study underscores the utility of combining transcriptomic data with metabolic modeling to uncover critical metabolic pathways that could serve as drug targets. Future studies could further investigate the role of L-valine in *K. pneumoniae* colonization and virulence or explore similar approaches for other resistant bacterial pathogens. Despite some limitations, such as potential biases from dataset selection and strain variability, this research demonstrates the power of systems biology approaches in identifying novel metabolic targets against bacterial pathogens.

In Silico Prediction of Potential Indigenous Microbial Biomarkers in **Penaeus vannamei** *Identified Through Meta-Analysis and Genome-Scale Metabolic Modelling [111]*

This is a machine-generated summary of:

Devika, Neelakantan Thulasi; Katneni, Vinaya Kumar; Jangam, Ashok Kumar; Suganya, Panjan Nathamuni; Shekhar, Mudagandur Shashi; Jithendran, Karingalakkandy Poochirian In Silico Prediction of Potential Indigenous Microbial Biomarkers in *Penaeus vannamei* Identified Through Meta-Analysis and Genome-Scale Metabolic Modelling [111]

Published in: Environmental Microbiome (2023)

Link to original: https://doi.org/10.1186/s40793-022-00458-6

If you want to cite the papers, please refer to the original.

For technical reasons we could not place the page where the original quote is coming from.

Research Highlights

With the advent of high-throughput sequencing, metagenome datasets have become increasingly accessible, facilitating extensive analysis of microbiome data for preventive and diagnostic purposes. In aquaculture, microbial consortia are crucial for improving the sustainability and productivity of aquatic organisms. However, the sector faces challenges from pathogenic microorganisms that threaten the health and yield of species like *Penaeus vannamei*, a widely cultured shrimp. Now, a new study by Neelakantan Thulasi Devika and colleagues integrates amplicon sequence data with genome-scale metabolic modeling to explore microbial interactions in *P. vannamei*, aiming to identify taxonomic biomarkers that could enhance shrimp health management.

The study conducted a meta-analysis using amplicon sequence data from the NCBI SRA database, focusing on *P. vannamei* samples associated with diseases like White Feces Disease and Acute Hepatopancreatic Necrotic Disease. The researchers processed single-end and paired-end sequences using the Quantitative Insights Into Microbial Ecology pipeline to assess microbial diversity and identify biomarkers. They employed constraint-based metabolic modeling to simulate microbial interactions under various nutrient environments, using tools like CarveMe for model reconstruction and CobraPy for flux balance analysis. This approach allowed them to predict growth rates and metabolic exchanges in microbial communities, identifying beneficial species that could limit pathogen growth.

The study advances our understanding of microbial interactions in aquaculture by identifying core genera associated with healthy and disease states in *P. vannamei*. Future studies could focus on experimental validation of these findings to develop probiotic formulations that leverage beneficial microorganisms for sustainable shrimp farming. The integration of amplicon sequencing with metabolic modeling offers a promising strategy for enhancing aquaculture health management by identifying native biomarkers that can improve water quality and control pathogenic species. This approach could be further refined with shotgun metagenomics to provide deeper insights into microbial community dynamics in aquaculture environments.

Metabolic Modeling Predicts Specific Gut Bacteria as Key Determinants for **Candida albicans** *Colonization Levels [112]*

This is a machine-generated summary of:

Mirhakkak, Mohammad H.; Schäuble, Sascha; Klassert, Tilman E.; Brunke, Sascha; Brandt, Philipp; Loos, Daniel; Uribe, Ruben V.; Senne de Oliveira Lino, Felipe; Ni, Yueqiong; Vylkova, Slavena; Slevogt, Hortense; Hube, Bernhard; Weiss, Glen J.; Sommer, Morten O. A.; Panagiotou, Gianni Metabolic Modeling Predicts Specific Gut Bacteria as Key Determinants for *Candida albicans* Colonization Levels [112]

Published in: The ISME Journal: Multidisciplinary Journal of Microbial Ecology (2020)

Link to original: https://doi.org/10.1038/s41396-020-00848-z

If you want to cite the papers, please refer to the original.

For technical reasons we could not place the page where the original quote is coming from.

Research Highlights

The fungus *Candida albicans* is a common commensal organism found on the mucosal surfaces of many healthy adults, but it can become pathogenic in immunocompromised individuals or under conditions of microbial dysbiosis. This opportunistic pathogen is responsible for a significant number of deaths worldwide, prompting increased research into the factors that influence its transition from a harmless commensal to a harmful pathogen. Recent studies have suggested that the gut microbiota plays a crucial role in this process, with certain microbial metabolic pathways potentially preserving immune function and inhibiting *C. albicans* growth. However, the specific dynamics and microbial contributors to these interactions remain poorly understood. Now, a new study by Mohammad H. Mirhakkak and colleagues has utilized genome-scale metabolic models (GSMMs) to explore the interactions between *C. albicans* and gut bacteria, providing insights into how these relationships might influence fungal fitness.

The researchers constructed a GSMM for *C. albicans* by starting with an automatically generated template model and enhancing it through manual curation and adaptation to phenotype microarray experiments. They used both existing data and new phenotypic microarray data from wild-type and mutant strains of *C. albicans*. The model's predictions were validated against other GSMMs for related species and further tested through in vitro experiments using defined media with specific metabolites predicted to affect fungal growth. Additionally, stool samples from human subjects were analyzed using metagenomics and internal transcribed spacer sequencing to identify bacterial species that significantly impact *C. albicans* metabolism and growth. The study also included experiments with bacterial spent media to assess its effects on fungal growth, providing a comprehensive analysis of the metabolic interactions between *C. albicans* and gut bacteria.

The study advances our understanding of how specific gut bacteria can modulate *C. albicans* growth through metabolic interactions, suggesting potential strategies for controlling fungal overgrowth by manipulating the gut microbiome. Future studies could focus on refining these models by incorporating additional dietary compositions or exploring multispecies interactions beyond pairwise analyses to better reflect in vivo conditions. Moreover, integrating host factors into these models could provide a more holistic view of the complex interplay between host, bacteria, and fungus in influencing *C. albicans* colonization levels. This research lays the groundwork for developing targeted probiotic or therapeutic interventions aimed at maintaining healthy microbial balance and preventing opportunistic infections by *C. albicans*.

Dynamic Fluctuations in a Bacterial Metabolic Network [113]

This is a machine-generated summary of:

Bi, Shuangyu; Kargeti, Manika; Colin, Remy; Farke, Niklas; Link, Hannes; Sourjik, Victor Dynamic Fluctuations in a Bacterial Metabolic Network [113]

Published in: Nature Communications (2023)

Link to original: https://doi.org/10.1038/s41467-023-37957-0

If you want to cite the papers, please refer to the original.

For technical reasons we could not place the page where the original quote is coming from.

Research Highlights

Metabolism is a fundamental aspect of cellular life, with metabolic pathways being highly conserved across species. These pathways not only fuel cellular activities but also regulate various cellular processes. Despite the deterministic view of central metabolism, cells operate as stochastic systems, leading to dynamic fluctuations in metabolic networks. While stochastic gene expression has been extensively studied, the origins and effects of posttranslational fluctuations remain less understood. Now, a new study by Shuangyu Bi and colleagues explores these fluctuations by examining pyruvate dynamics in single *Escherichia coli* cells using a FRET-based sensor.

The study employed a kinetic model of *E. coli* glycolysis to predict oscillations in pyruvate levels upon changes in glucose uptake. The model included key allosteric interactions and was tested across various parameter sets to identify conditions that favor oscillations. Experimentally, the researchers used single-cell FRET microscopy to monitor pyruvate levels in *E. coli* cells exposed to different carbon sources. They observed significant fluctuations in pyruvate levels on the timescale of several minutes when cells were stimulated with glucose or other glycolytic substrates. These fluctuations were linked to specific biochemical reactions and regulatory processes within the cell, suggesting a complex interplay between metabolic pathways.

The findings highlight the potential for metabolic networks to exhibit dynamic behavior beyond deterministic models. The study suggests that these fluctuations may be influenced by multiple factors, including allosteric regulation and energy state of the cell. Future studies could explore the physiological significance of these fluctuations and their impact on bacterial metabolism and gene regulation. Understanding these dynamics could provide insights into how cells adapt to changing environments and maintain homeostasis despite inherent stochasticity in metabolic processes.

A Systems Biology Framework for Modeling Metabolic Enzyme Inhibition of Mycobacterium tuberculosis *[114]*

This is a machine-generated summary of:

Fang, Xin; Wallqvist, Anders; Reifman, Jaques A Systems Biology Framework for Modeling Metabolic Enzyme Inhibition of *Mycobacterium tuberculosis* [114]

Published in: BMC Systems Biology (2009)
Link to original: https://doi.org/10.1186/1752-0509-3-92

If you want to cite the papers, please refer to the original.
For technical reasons we could not place the page where the original quote is coming from.

Research Highlights
System-level networks of biological processes, particularly metabolic networks, are crucial for understanding the phenotype of an organism. These networks consist of numerous interconnected chemical reactions that enable cellular function and growth. While metabolic network assemblies exist for several bacterial species, including *Mycobacterium tuberculosis* (*M. tuberculosis*), current models often lack integration of enzyme inhibition kinetics, flux balance analysis (FBA), and cell growth dynamics. Now, a new study by Xin Fang and colleagues presents a comprehensive framework that combines these three components to model the growth inhibition of *M. tuberculosis*, the causative agent of tuberculosis.

The study developed a mathematical framework that integrates enzyme inhibition kinetics, metabolic network simulation, and cell growth dynamics to quantitatively model drug inhibition effects on bacterial growth. The researchers used this framework to simulate the effects of two metabolic inhibitors, 3-nitropropionate (3-NP) and sAMS, on *M. tuberculosis* cells in vitro. The framework connects how an inhibitor affects metabolic reaction fluxes, how these changes impact organism growth rates, and ultimately how they influence bacterial cell concentrations. By employing this integrated approach, the team was able to generate dose–response curves that closely matched experimental data from independent studies. The framework's ability to predict these curves demonstrates its potential for understanding the impact of metabolic inhibitors on bacterial growth.

The advances made by Fang and colleagues provide a significant step forward in modeling complex interactions within metabolic networks under drug influence. This integrated approach offers insights into potential therapeutic targets for tuberculosis treatment by simulating inhibitor effects on key metabolic pathways. Future study could expand this framework to include additional cellular processes such as gene expression changes or adaptive responses to drug exposure. Moreover, coupling this model with other systems biology approaches could enhance its applicability in more complex in vivo environments, potentially leading to more effective strategies for combating drug-resistant strains of *M. tuberculosis*.

A Catalog of the Diversity and Ubiquity of Bacterial Microcompartments [115]

This is a machine-generated summary of:

Sutter, Markus; Melnicki, Matthew R.; Schulz, Frederik; Woyke, Tanja; Kerfeld, Cheryl A. A Catalog of the Diversity and Ubiquity of Bacterial Microcompartments [115]

Published in: Nature Communications (2021)

Link to original: https://doi.org/10.1038/s41467-021-24126-4

If you want to cite the papers, please refer to the original.

For technical reasons we could not place the page where the original quote is coming from.

Research Highlights

Bacterial microcompartments (BMCs) are protein-based metabolic organelles that enhance metabolic processes by encapsulating enzymes and substrates within a modular shell. Initially discovered in cyanobacteria as carboxysomes, BMCs have since been identified in various bacterial phyla, performing diverse catabolic functions. Despite advances in genomic sequencing, the full extent of BMC diversity and distribution remains underexplored. Now, a new study by Markus Sutter and colleagues has significantly expanded our understanding of BMCs by analyzing extensive genomic datasets to identify over 6,000 BMC loci, revealing a doubling of known functional diversity and distribution across bacterial phyla.

The study utilized a comprehensive bioinformatic approach to mine recent genomic and metagenomic datasets, compiling a database of BMC loci. The researchers employed phylogenomic methods to classify shell proteins into distinct types and subtypes, allowing for the identification of new functional BMC types. By constructing phylogenetic trees for shell protein sequences and developing profile Hidden Markov Models (HMMs), the team was able to cluster loci into distinct BMC types based on shell protein composition and enzymatic components. This approach enabled the discovery of new BMC types with unique functional characteristics, such as those involved in sugar-phosphate utilization and aromatic compound degradation. The study also highlighted the modular nature of BMC shells, which are composed of various hexagonal building blocks with different permselectivities.

The findings from this study underscore the widespread occurrence and functional diversity of BMCs across bacterial phyla, emphasizing their role in metabolic flexibility and adaptation to diverse environments. Future studies could focus on elucidating the specific functions of newly identified BMC types through genetic and biochemical analyses. Additionally, the insights gained from this research could inform metabolic engineering strategies by leveraging the modular construction of BMC shells for synthetic biology applications. The expanded catalog of BMCs provides a valuable resource for understanding microbial metabolism and its evolutionary history, with potential implications for biotechnology and environmental science.

9.3 Conclusion

The studies presented in this chapter collectively illustrate the transformative potential of metabolic modeling in understanding and combating bacterial infections. By integrating computational and mathematical frameworks with experimental data, researchers can simulate complex metabolic networks, predict physiological responses, and identify novel therapeutic targets. These approaches offer a systems-level view of microbial behavior, enabling precise interventions in both clinical and environmental contexts.

One of the most significant contributions of metabolic modeling is its ability to enhance metabolite annotation and profiling. The PeakDecoder algorithm, as demonstrated by Bilbao et al., exemplifies how machine learning can improve the accuracy and throughput of metabolomics. By combining LC-IM-MS with targeted data extraction and decoy generation, PeakDecoder enables confident identification of metabolites in engineered strains. This innovation accelerates synthetic biology workflows and supports the development of microbial platforms for biomanufacturing and environmental applications.

Genome-scale metabolic models (GSMMs) are foundational tools in systems biology. Chung et al. highlight their utility in identifying antimicrobial targets in multidrug-resistant Gram-negative pathogens. GSMMs integrate multi-omics data to simulate bacterial physiology and predict drug responses, offering a rational framework for drug discovery. The reconstruction process, involving tools like Model SEED and CarveMe, ensures biological fidelity and computational robustness. These models pave the way for designing synergistic therapies and understanding resistance mechanisms at a systems level.

The study by Jenior et al. demonstrates how GSMMs can reveal context-specific metabolic adaptations in clinical isolates of *Klebsiella pneumoniae*. Increased reliance on valine catabolism in infection settings suggests a metabolic vulnerability that could be exploited therapeutically. This approach underscores the importance of integrating transcriptomic data with metabolic modeling to uncover traits relevant to pathogenesis and drug resistance.

In aquaculture, metabolic modeling supports health management and sustainability. Devika et al. use GSMMs to identify microbial biomarkers in *Penaeus vannamei*, predicting beneficial interactions that limit pathogen growth. This integrative strategy informs probiotic development and enhances disease prevention in shrimp farming, demonstrating the versatility of metabolic modeling beyond human health.

Host–microbe interactions are another frontier for metabolic modeling. Mirhakkak et al. construct a GSMM for *Candida albicans* to explore its metabolic interplay with gut bacteria. Their findings reveal that specific bacterial species can modulate fungal growth through metabolite exchange, influencing colonization and pathogenicity. This research lays the groundwork for microbiome-targeted therapies to prevent opportunistic infections.

Dynamic fluctuations in bacterial metabolism challenge traditional deterministic models. Bi et al. investigate pyruvate oscillations in *E. coli*, revealing stochastic behavior influenced by allosteric regulation and energy state. These insights highlight the complexity of metabolic control and suggest that variability in metabolic networks may play a role in bacterial adaptability and resilience.

In the fight against tuberculosis, Fang et al. present a systems biology framework that integrates enzyme inhibition kinetics with metabolic modeling. Their model accurately predicts dose–response curves for metabolic inhibitors, offering a quantitative tool for evaluating drug efficacy and identifying therapeutic targets. This approach enhances our ability to simulate drug effects and design effective treatments for persistent infections.

The expanded catalog of bacterial microcompartments (BMCs) by Sutter et al. reveals the diversity and ubiquity of these protein-based organelles. BMCs encapsulate metabolic enzymes and substrates, enhancing efficiency and specificity. Their modular architecture and functional diversity offer exciting opportunities for metabolic engineering and synthetic biology, enabling the design of tailored microbial systems for biotechnological applications.

In conclusion, metabolic modeling provides a powerful lens through which to examine bacterial physiology, pathogenesis, and therapeutic intervention. The studies in this chapter demonstrate that computational and mathematical approaches are essential for unraveling the complexities of microbial metabolism and guiding the development of innovative solutions in medicine, biotechnology, and environmental science. As tools and datasets continue to evolve, metabolic modeling will play an increasingly central role in shaping the future of microbiological research and public health.

Bibliography

1. Tsugawa, H. et al. A lipidome atlas in MS-DIAL 4. *Nat. Biotechnol. 38*, 1159–1163 (2020).
2. Li, Y. et al. Spectral entropy outperforms MS/MS dot product similarity for small-molecule compound identification. *Nat. Methods 18*, 1524–1531 (2021).
3. Coradetti, S. T. et al. Functional genomics of lipid metabolism in the oleaginous yeast *Rhodosporidium toruloides*. *Elife 7*, e32110 (2018).
4. Cox, J. & Mann, M. MaxQuant enables high peptide identification rates, individualized p.p.b.-range mass accuracies and proteome-wide protein quantification. *Nat. Biotechnol. 26*, 1367–1372 (2008).
5. World Health Organization. Antimicrobial resistance: global report on surveillance 2014. 2016:257. https://apps.who.int/iris/bitstream/handle/10665/112642/9789241564748_eng.pdf?sequence=1.
6. World Health Organization. Global strategy for containment of antimicrobial resistance. 2001. https://apps.who.int/iris/bitstream/handle/10665/66860/WHO_CDS_CSR_DRS_2001.2.pdf?sequence=1.
7. Gu C, Kim GB, Kim WJ, Kim HU, Lee SY. Current status and applications of genome-scale metabolic models. Genome Biol. 2019;20. https://doi.org/10.1186/s13059-019-1730-3.
8. Mobegi FM, van Hijum SAFT, Burghout P, Bootsma HJ, de Vries SPW, van der Gaast-de Jongh CE, et al. From microbial gene essentiality to novel antimicrobial drug targets. BMC Genom. 2014;15:958. https://doi.org/10.1186/1471-2164-15-958.
9. Fields FR, Lee SW, McConnell MJ. Using bacterial genomes and essential genes for the development of new antibiotics. Biochem Pharmacol. 2017;134:74–86. https://doi.org/10.1016/j.bcp.2016.12.002.
10. Moir DT, Shaw KJ, Hare RS, Vovis GF. Genomics and antimicrobial drug discovery. Antimicrob Agents Chemother. 1999;43:439–46. https://doi.org/10.1128/aac.43.3.439.
11. Oberhardt, M. A., Puchalka, J., Fryer, K. E., dos Santos, V. A. P. M. & Papin, J. A. Genome-scale metabolic network analysis of the opportunistic pathogen *Pseudomonas aeruginosa* PAO1. *J. Bacteriol. 190*, 2790–2803 (2008).
12. Oberhardt, M. A., Puchalka, J., dos Santos, V. A. P. M. & Papin, J. A. Reconciliation of genome-scale metabolic reconstructions for comparative systems analysis. *PLoS Comput. Biol. 7*, e1001116 (2011).
13. Bartell JA, Blazier AS, Yen P, Thøgersen JC, Jelsbak L, Goldberg JB, et al. Reconstruction of the metabolic network of *Pseudomonas aeruginosa* to interrogate virulence factor synthesis. Nat Commun. 2017;8. https://doi.org/10.1038/ncomms14631.
14. Zhu Y, Czauderna T, Zhao J, Klapperstueck M, Maifiah MHM, Han ML, et al. Genome-scale metabolic modeling of responses to polymyxins in *Pseudomonas aeruginosa*. Gigascience. 2018;7:1–18. https://doi.org/10.1093/gigascience/giy021.
15. Presta L, Bosi E, Mansouri L, Dijkshoorn L, Fani R, Fondi M. Constraint-based modeling identifies new putative targets to fight colistin-resistant *A. baumannii* infections. Sci Rep. 2017;7. https://doi.org/10.1038/s41598-017-03416-2.
16. Norsigian CJ, Kavvas E, Seif Y, Palsson BO, Monk JM. iCN718, an updated and improved genome-scale metabolic network reconstruction of *Acinetobacter baumannii* AYE. Front Genet. 2018;9. https://doi.org/10.3389/fgene.2018.00121.
17. Zhu Y, Zhao J, Maifiah MHM, Velkov T, Schreiber F, Li J. Metabolic responses to polymyxin treatment in *Acinetobacter baumannii* ATCC 19606: integrating transcriptomics and metabolomics with genome-scale metabolic modeling. mSystems. 2019;4:e00157–18. https://doi.org/10.1128/msystems.00157-18.
18. Liao YC, Huang TW, Chen FC, Charusanti P, Hong JSJ, Chang HY, et al. An experimentally validated genome-scale metabolic reconstruction of *Klebsiella pneumoniae* MGH 78578, iYL1228. J Bacteriol. 2011;193:1710–7. https://doi.org/10.1128/JB.01218-10.

19. Henry CS, Rotman E, Lathem WW, Tyo KEJ, Hauser AR, Mandel MJ. Generation and validation of the iKp1289 Metabolic model for *Klebsiella pneumoniae* KPPR1. J Infect Dis. 2017;215:S37–43. https://doi.org/10.1093/infdis/jiw465.
20. Ramos PIP, Fernández Do Porto D, Lanzarotti E, Sosa EJ, Burguener G, Pardo AM, et al. An integrative, multi-omics approach towards the prioritization of Klebsiella pneumoniae drug targets. Sci Rep. 2018;8. https://doi.org/10.1038/s41598-018-28916-7.
21. Norsigian CJ, Attia H, Szubin R, Yassin AS, Palsson B, Aziz RK, et al. Comparative genome-scale metabolic modeling of metallo-beta-lactamase-producing multidrug-resistant *Klebsiella pneumoniae* clinical isolates. Front Cell Infect Microbiol. 2019;9. https://doi.org/10.3389/fcimb.2019.00161.
22. Dunphy LJ, Papin JA. Biomedical applications of genome-scale metabolic network reconstructions of human pathogens. Curr Opin Biotechnol. 2018;51:70–9. https://doi.org/10.1016/j.copbio.2017.11.014.
23. Chavali AK, D'Auria KM, Hewlett EL, Pearson RD, Papin JA. A metabolic network approach for the identification and prioritization of antimicrobial drug targets. Trends Microbiol. 2012;20:113–23. https://doi.org/10.1016/j.tim.2011.12.004.
24. Karp PD, Paley S, Romero P. The pathway tools software. Bioinformatics. 2002;18:S225–32. https://doi.org/10.1093/bioinformatics/18.suppl_1.S225.
25. Pinney JW, Shirley MW, McConkey GA, Westhead DR. metaSHARK: software for automated metabolic network prediction from DNA sequence and its application to the genomes of *Plasmodium falciparum* and *Eimeria tenella*. Nucleic Acids Res. 2005;33:1399–409. https://doi.org/10.1093/nar/gki285.
26. Thiele, I. & Palsson, B. Ø. A protocol for generating a high-quality genome-scale metabolic reconstruction. *Nat. Protoc. 5*, 93–121 (2010).
27. Henry CS, DeJongh M, Best AA, Frybarger PM, Linsay B, Stevens RL. High-throughput generation, optimization and analysis of genome-scale metabolic models. Nat Biotechnol. 2010;28(9):977–U922. https://doi.org/10.1038/nbt.1672.
28. Wang H, Marcišauskas S, Sánchez BJ, Domenzain I, Hermansson D, Agren R, et al. RAVEN 2.0: a versatile toolbox for metabolic network reconstruction and a case study on *Streptomyces coelicolor*. PLoS Comput Biol. 2018;14:e1006541 https://doi.org/10.1371/journal.pcbi.1006541.
29. Dias O, Rocha M, Ferreira EC, Rocha I. Reconstructing genome-scale metabolic models with merlin. Nucleic Acids Res. 2015;43:3899–910.
30. Aite M, Chevallier M, Frioux C, Trottier C, Got J, Cortés MP, et al. Traceability, reproducibility and wiki-exploration for "à-la-carte" reconstructions of genome-scale metabolic models. PLoS Comput Biol. 2018;14:e1006146. https://doi.org/10.1371/journal.pcbi.1006146.
31. Machado D, Andrejev S, Tramontano M, Patil KR. Fast automated reconstruction of genome-scale metabolic models for microbial species and communities. Nucleic Acids Res. 2018;46:7542–53. https://doi.org/10.1093/nar/gky537.
32. Hanemaaijer M, Olivier BG, Röling WFM, Bruggeman FJ, Teusink B. Model-based quantification of metabolic interactions from dynamic microbial-community data. PLoS ONE. 2017;12:e0173183. https://doi.org/10.1371/journal.pone.0173183.
33. Mendoza SN, Olivier BG, Molenaar D, Teusink B. A systematic assessment of current genome-scale metabolic reconstruction tools. Genome Biol. 2019;20. https://doi.org/10.1186/s13059-019-1769-1.
34. Ling ML, Apisarnthanarak A, Madriaga G. The burden of healthcare-associated infections in southeast Asia: a systematic literature review and meta-analysis. Clin Infect Dis. 2015;60:1690–9. https://doi.org/10.1093/cid/civ095.
35. Exner M, Bhattacharya S, Christiansen B, Gebel J, Goroncy-Bermes P, Hartemann P, et al. Antibiotic resistance: what is so special about multidrug-resistant Gram-negative bacteria? GMS Hyg Infect Control. 2017;12. https://doi.org/10.3205/dgkh000290.

36. Cesur MF, Siraj B, Uddin R, Durmuş S, Çakır T. Network-based metabolism-centered screening of potential drug targets in *Klebsiella pneumoniae* at genome scale. Front Cell Infect Microbiol. 2020;9. https://doi.org/10.3389/fcimb.2019.00447.
37. Kim HU, Kim TY, Lee SY. Genome-scale metabolic network analysis and drug targeting of multi-drug resistant pathogen *Acinetobacter baumannii* AYE. Mol Biosyst. 2010;6:339–48. https://doi.org/10.1039/b916446d.
38. Vital-Lopez FG, Reifman J, Wallqvist A. Biofilm formation mechanisms of *Pseudomonas aeruginosa* predicted via genome-scale kinetic models of bacterial metabolism. PLoS Comput Biol. 2015;11:e1004452. https://doi.org/10.1371/journal.pcbi.1004452.
39. Biggs MB, Papin JA. Novel multiscale modeling tool applied to *Pseudomonas aeruginosa* biofilm formation. PLoS ONE. 2013;8:e78011. https://doi.org/10.1371/journal.pone.0078011.
40. Stokes, J. M., Lopatkin, A. J., Lobritz, M. A. & Collins, J. J. Bacterial metabolism and antibiotic efficacy. *Cell Metab. 30*, 251–259 (2019).
41. Hasan, S., Daugelat, S., Rao, P. S. S. & Schreiber, M. Prioritizing genomic drug targets in pathogens: application to Mycobacterium tuberculosis. *PLoS Comput. Biol. 2*, e61 (2006).
42. Song, J.-H. & Ko, K. S. In *Microbial Gene Essentiality: Protocols and Bioinformatics* (eds. Osterman, A. L. & Gerdes, S. Y.) vol. 416, 401–408 (Humana Press, 2008).
43. Ramos, P. I. P. et al. An integrative, multi-omics approach towards the prioritization of Klebsiella pneumoniae drug targets. *Sci. Rep. 8*, 10755 (2018).
44. Hadizadeh, M. et al. Genome-wide identification of potential drug target in *Enterobacteriaceae* family: a homology-based method. *Microb. Drug Resist. 24*, 8–17 (2018).
45. Lee, M. et al. Network integrative genomic and transcriptomic analysis of carbapenem-resistant *Klebsiella pneumoniae* strains identifies genes for antibiotic resistance and virulence. *mSystems 4*, e00202–e00219 (2019).
46. Long, D. et al. Phenotypical profile and global transcriptomic profile of Hypervirulent Klebsiella pneumoniae due to carbapenemase-encoding plasmid acquisition. *BMC Genomics 20*, 480 (2019).
47. Guilhen, C. et al. Transcriptional profiling of Klebsiella pneumoniae defines signatures for planktonic, sessile and biofilm-dispersed cells. *BMC Genomics 17*, 237 (2016).
48. Bruchmann, S. et al. Deep transcriptome profiling of clinical *K lebsiella pneumoniae* isolates reveals strain and sequence type-specific adaptation: comparative transcriptomics of *Klebsiella pneumoniae*. *Environ. Microbiol. 17*, 4690–4710 (2015).
49. Liao, Y. C. et al. *Model: iYL1228 Klebsiella pneumoniae subsp. pneumoniae MGH 78578.* (BiGG Model Database, 2019).
50. Gu, C., Kim, G. B., Kim, W. J., Kim, H. U. & Lee, S. Y. Current status and applications of genome-scale metabolic models. *Genome Biol. 20*, 121 (2019).
51. Ruppin, E., Papin, J. A., de Figueiredo, L. F. & Schuster, S. Metabolic reconstruction, constraint-based analysis and game theory to probe genome-scale metabolic networks. *Curr. Opin. Biotechnol. 21*, 502–510 (2010).
52. Jenior, M. L., Moutinho, T. J., Dougherty, B. V. & Papin, J. A. Transcriptome-guided parsimonious flux analysis improves predictions with metabolic networks in complex environments. *PLoS Comput. Biol. 16*, e1007099 (2020).
53. Liao, Y.-C. et al. An experimentally validated genome-scale metabolic reconstruction of *Klebsiella pneumoniae* MGH 78578, *i*YL1228. *J. Bacteriol. 193*, 1710–1717 (2011).
54. Bachman, M. A. et al. Genome-wide identification of Klebsiella pneumoniae fitness genes during lung infection. *mBio 6*, e00775 (2015).
55. Liu, S. et al. Metabolic regulation protects mice against Klebsiella pneumoniae lung infection. *Exp. Lung Res. 44*, 301–311 (2018).
56. Ebrahim, A., Lerman, J. A., Palsson, B. O. & Hyduke, D. R. COBRApy: COnstraints-Based Reconstruction and Analysis for Python. *BMC Syst. Biol. 7*, 74 (2013).
57. Opdam, S. et al. A systematic evaluation of methods for tailoring genome-scale metabolic models. *Cell Syst. 4*, 318–329.e6 (2017).

58. Cheng Y, Ge C, Li W, Yao H. The intestinal bacterial community and functional potential of *Litopenaeus vannamei* in the coastal areas of china. Microorganisms. 2021. https://doi.org/10.3390/microorganisms9091793.
59. Zheng Y, Yu M, Liu J, Qiao Y, Wang L, Li Z, Zhang XH, Yu M. Bacterial community associated with healthy and diseased Pacific white shrimp (*Litopenaeus vannamei*) larvae and rearing water across different growth stages. Front Microbiol. 2017;8(JUL):1–11. https://doi.org/10.3389/fmicb.2017.01362.
60. Shao J, Jiang K, Wang L. *Litopenaeus vannamei* fed diets with different replacement levels of fish meal by fish silage: a molecular approach on intestinal microbiota. Aquac Nutr. 2019;25(3):721–8. https://doi.org/10.1111/anu.12900.
61. Devika NT, Jangam AK, Katneni VK, Patil PK, Nathamuni S, Shekhar MS. In silico prediction of novel probiotic species limiting pathogenic *Vibrio* growth using constraint-based genome scale metabolic modeling. Front Cell Infect Microbiol. 2021;11(September):1–10. https://doi.org/10.3389/fcimb.2021.752477.
62. Ibrahim M, Raman K. Two-species community design of lactic acid bacteria for optimal production of lactate. Comput Struct Biotechnol J. 2021;19:6039–49. https://doi.org/10.1016/j.csbj.2021.11.009.
63. Thommes M, Wang T, Zhao Q, Paschalidis IC, Segrè D. Designing metabolic division of labor in microbial communities. MSystems. 2019;4(2):1–21. https://doi.org/10.1128/msystems.00263-18.
64. Machado D, Maistrenko OM, Andrejev S, Kim Y, Bork P, Patil KR, Patil KR. Polarization of microbial communities between competitive and cooperative metabolism. Nat Ecol Evol. 2021. https://doi.org/10.1038/s41559-020-01353-4.
65. Zorrilla F, Buric F, Patil KR, Zelezniak A. MetaGEM: reconstruction of genome scale metabolic models directly from metagenomes. Nucleic Acids Res. 2021. https://doi.org/10.1093/nar/gkab815.
66. Gao S, Pan L, Huang F, Song M, Tian C, Zhang M. Metagenomic insights into the structure and function of intestinal microbiota of the farmed Pacific white shrimp (*Litopenaeus vannamei*). Aquaculture. 2019;499(April 2018):109–18. https://doi.org/10.1016/j.aquaculture.2018.09.026.
67. Förster TM, Mogavero S, Dräger A, Graf K, Polke M, Jacobsen ID, et al. Enemies and brothers in arms: Candida albicans and gram-positive bacteria. Cell Microbiol. 2016;18:1709–15.
68. da Silva Dantas A, Lee KK, Raziunaite I, Schaefer K, Wagener J, Yadav B, et al. Cell biology of Candida albicans–host interactions. Curr Opin Microbiol. 2016;34:111–8.
69. Kaewsrichan J, Peeyananjarassri K, Kongprasertkit J. Selection and identification of anaerobic lactobacilli producing inhibitory compounds against vaginal pathogens. FEMS Immunol Med Microbiol. 2006;48:75–83.
70. Feist AM, Henry CS, Reed JL, Krummenacker M, Joyce AR, Karp PD, et al. A genome-scale metabolic reconstruction for Escherichia coli K-12 MG1655 that accounts for 1260 ORFs and thermodynamic information. Mol Syst Biol. 2007;3:121.
71. Feist AM, Palsson BØ. The growing scope of applications of genome-scale metabolic reconstructions using Escherichia coli. Nat Biotechnol. 2008;26:659–67.
72. Trinh CT, Srienc F. Metabolic engineering of Escherichia coli for efficient conversion of glycerol to ethanol. Appl Environ Microbiol. 2009;75:6696–705.
73. Gomes de Oliveira Dal'Molin C, Quek L-E, Saa PA, Nielsen LK. A multi-tissue genome-scale metabolic modeling framework for the analysis of whole plant systems. Front Plant Sci. 2015;6:4.
74. van Hoek MJA, Merks RMH. Redox balance is key to explaining full vs. partial switching to low-yield metabolism. BMC Syst Biol. 2012;6:22.
75. Pfeiffer T, Morley A. An evolutionary perspective on the Crabtree effect. Front Mol Biosci. 2014;1:17.
76. Nilsson A, Nielsen J. Metabolic trade-offs in yeast are caused by F1F0-ATP synthase. Sci Rep. 2016;6:22264.

77. Shlomi T, Benyamini T, Gottlieb E, Sharan R, Ruppin E. Genome-scale metabolic modeling elucidates the role of proliferative adaptation in causing the Warburg effect. PLoS Comput Biol. 2011;7:e1002018.
78. Eldar, A. & Elowitz, M. B. Functional roles for noise in genetic circuits. *Nature 467*, 167–173 (2010).
79. Farquhar, K. S. et al. Role of network-mediated stochasticity in mammalian drug resistance. *Nat. Commun. 10*, 2766 (2019).
80. Levine, E. & Hwa, T. Stochastic fluctuations in metabolic pathways. *Proc. Natl Acad. Sci.104*, 9224–9229 (2007).
81. Rao, C. V., Wolf, D. M. & Arkin, A. P. Control, exploitation and tolerance of intracellular noise. *Nature 420*, 231–237 (2002).
82. Schwabe, A., Dobrzynski, M., Rybakova, K., Verschure, P. & Bruggeman, F. J. Origins of stochastic intracellular processes and consequences for cell-to-cell variability and cellular survival strategies. *Methods Enzymol. 500*, 597–625 (2011).
83. Silva-Rocha, R. & de Lorenzo, V. Noise and robustness in prokaryotic regulatory networks. *Annu. Rev. Microbiol. 64*, 257–275 (2010).
84. Takhaveev, V. & Heinemann, M. Metabolic heterogeneity in clonal microbial populations. *Curr. Opin. Microbiol. 45*, 30–38 (2018).
85. Tsimring, L. S. Noise in biology. *Rep. Prog. Phys. 77*, 026601 (2014).
86. Co, A. D., Lagomarsino, M. C., Caselle, M. & Osella, M. Stochastic timing in gene expression for simple regulatory strategies. *Nucl. Acids Res. 45*, 1069–1078 (2017).
87. Desai, R. V. et al. A DNA repair pathway can regulate transcriptional noise to promote cell fate transitions. *Science 373*, eabc6506 (2021).
88. Elowitz, M. B., Levine, A. J., Siggia, E. D. & Swain, P. S. Stochastic gene expression in a single cell. *Science 297*, 1183–1186 (2002).
89. Kiviet, D. J. et al. Stochasticity of metabolism and growth at the single-cell level. *Nature 514*, 376–379 (2014).
90. Scott, M., Gunderson, C. W., Mateescu, E. M., Zhang, Z. & Hwa, T. Interdependence of cell growth and gene expression: Origins and consequences. *Sci. (80-.) 330*, 1099–1102 (2010).
91. Cerulus, B., New, A. M., Pougach, K. & Verstrepen, K. J. Noise and epigenetic inheritance of single-cell division times influence population fitness. *Curr. Biol. 26*, 1138–1147 (2016).
92. Nikolic, N., Barner, T. & Ackermann, M. Analysis of fluorescent reporters indicates heterogeneity in glucose uptake and utilization in clonal bacterial populations. *BMC Microbiol 13*, 258 (2013).
93. Dunlop, M. J., Cox, R. S. 3rd, Levin, J. H., Murray, R. M. & Elowitz, M. B. Regulatory activity revealed by dynamic correlations in gene expression noise. *Nat. Genet. 40*, 1493–1498 (2008).
94. Kotte, O., Volkmer, B., Radzikowski, J. L. & Heinemann, M. Phenotypic bistability in *Escherichia coli*'s central carbon metabolism. *Mol. Syst. Biol. 10*, 736 (2014).
95. Richard, P. The rhythm of yeast. *FEMS Microbiol. Rev. 27*, 547–557 (2003).
96. Merrins, M. J., Van Dyke, A. R., Mapp, A. K., Rizzo, M. A. & Satin, L. S. Direct measurements of oscillatory glycolysis in pancreatic islet beta-cells using novel fluorescence resonance energy transfer (FRET) biosensors for pyruvate kinase M2 activity. *J. Biol. Chem. 288*, 33312–33322 (2013).
97. Somavanshi, R., Ghosh, B. & Sourjik, V. Sugar influx sensing by the phosphotransferase system of *Escherichia coli*. *PLoS Biol. 14*, e2000074 (2016).
98. Berg, S. et al. ilastik: interactive machine learning for (bio)image analysis. *Nat. Methods. 16*, 1226–1232 (2019).
99. Suchanek, V. M. et al. Chemotaxis and cyclic-di-GMP signalling control surface attachment of *Escherichia coli*. *Mol. Microbiol 113*, 728–739 (2020).
100. Muñoz-Elías, E. J. & McKinney, J. D. *Mycobacterium tuberculosis* isocitrate lyases 1 and 2 are jointly required for *in vivo* growth and virulence. *Nature Med. 11*, 638–644 (2005).

101. Ferreras JA, Ryu JS, Di Lello F, Tan DS, Quadri LE: Small-molecule inhibition of siderophore biosynthesis in Mycobacterium tuberculosis and Yersinia pestis. Nat Chem Biol. 2005, 1: 29-32. https://doi.org/10.1038/nchembio706
102. Drews, G. & Niklowitz, W. Beiträge zur Cytologie der Blaualgen. II. Zentroplasma und granuläre Einschlüsse von Phormidium uncinatum. *Arch. f.ür. Mikrobiologie 24*, 147–162 (1956).
103. Axen, S. D., Erbilgin, O. & Kerfeld, C. A. A taxonomy of bacterial microcompartment loci constructed by a novel scoring method. *PLoS Comput. Biol. 10*, e1003898 (2014).
104. Jorda, J., Lopez, D., Wheatley, N. M. & Yeates, T. O. Using comparative genomics to uncover new kinds of protein-based metabolic organelles in bacteria. *Protein Sci. 22*, 179–195 (2013).
105. Cai, F., Sutter, M., Bernstein, S. L., Kinney, J. N. & Kerfeld, C. A. Engineering bacterial microcompartment shells: chimeric shell proteins and chimeric carboxysome shells. *ACS Synth. Biol. 4*, 444–453 (2015).
106. Slininger Lee, M. F., Jakobson, C. M. & Tullman-Ercek, D. Evidence for improved encapsulated pathway behavior in a bacterial microcompartment through shell protein engineering. *ACS Synthetic Biol.* https://doi.org/10.1021/acssynbio.7b00042 (2017).
107. Castelle, C. J. & Banfield, J. F. Major new microbial groups expand diversity and alter our understanding of the tree of life. *Cell 172*, 1181–1197 (2018).
108. Bilbao, Aivett; Munoz, Nathalie; Kim, Joonhoon; Orton, Daniel J.; Gao, Yuqian; Poorey, Kunal; Pomraning, Kyle R.; Weitz, Karl; Burnet, Meagan; Nicora, Carrie D.; Wilton, Rosemarie; Deng, Shuang; Dai, Ziyu; Oksen, Ethan; Gee, Aaron; Fasani, Rick A.; Tsalenko, Anya; Tanjore, Deepti; Gardner, James; Smith, Richard D.; Michener, Joshua K.; Gladden, John M.; Baker, Erin S.; Petzold, Christopher J.; Kim, Young-Mo; Apffel, Alex; Magnuson, Jon K.; Burnum-Johnson, Kristin E. PeakDecoder enables machine learning-based metabolite annotation and accurate profiling in multidimensional mass spectrometry measurements. Nature Communications (2023). https://doi.org/10.1038/s41467-023-37031-9
109. Chung, Wan Yean; Zhu, Yan; Mahamad Maifiah, Mohd Hafidz; Shivashekaregowda, Naveen Kumar Hawala; Wong, Eng Hwa; Abdul Rahim, Nusaibah Novel antimicrobial development using genome-scale metabolic model of Gram-negative pathogens: a review. The Journal of Antibiotics (2020). https://doi.org/10.1038/s41429-020-00366-2
110. Jenior, Matthew L.; Dickenson, Mary E.; Papin, Jason A. Genome-scale metabolic modeling reveals increased reliance on valine catabolism in clinical isolates of Klebsiella pneumoniae. npj Systems Biology and Applications (2022). https://doi.org/10.1038/s41540-022-00252-7
111. Devika, Neelakantan Thulasi; Katneni, Vinaya Kumar; Jangam, Ashok Kumar; Suganya, Panjan Nathamuni; Shekhar, Mudagandur Shashi; Jithendran, Karingalakkandy Poochirian In silico prediction of potential indigenous microbial biomarkers in Penaeus vannamei identified through meta-analysis and genome-scale metabolic modelling. Environmental Microbiome (2023). https://doi.org/10.1186/s40793-022-00458-6
112. Mirhakkak, Mohammad H.; Schäuble, Sascha; Klassert, Tilman E.; Brunke, Sascha; Brandt, Philipp; Loos, Daniel; Uribe, Ruben V.; Senne de Oliveira Lino, Felipe; Ni, Yueqiong; Vylkova, Slavena; Slevogt, Hortense; Hube, Bernhard; Weiss, Glen J.; Sommer, Morten O. A.; Panagiotou, Gianni Metabolic modeling predicts specific gut bacteria as key determinants for Candida albicans colonization levels. The ISME Journal: Multidisciplinary Journal of Microbial Ecology (2020). https://doi.org/10.1038/s41396-020-00848-z
113. Bi, Shuangyu; Kargeti, Manika; Colin, Remy; Farke, Niklas; Link, Hannes; Sourjik, Victor Dynamic fluctuations in a bacterial metabolic network. Nature Communications (2023). https://doi.org/10.1038/s41467-023-37957-0
114. Fang, Xin; Wallqvist, Anders; Reifman, Jaques A systems biology framework for modeling metabolic enzyme inhibition of Mycobacterium tuberculosis. BMC Systems Biology (2009). https://doi.org/10.1186/1752-0509-3-92
115. Sutter, Markus; Melnicki, Matthew R.; Schulz, Frederik; Woyke, Tanja; Kerfeld, Cheryl A. A catalog of the diversity and ubiquity of bacterial microcompartments. Nature Communications (2021). https://doi.org/10.1038/s41467-021-24126-4

Chapter 10
Metabolic Engineering and Novel Therapeutics

Vijay Soni and Aditya Upadhyay

10.1 Introduction by the Editor

Metabolic engineering has emerged as a transformative discipline in microbiology and biotechnology, enabling the design and optimization of microbial systems for therapeutic, industrial, and environmental applications. In the context of bacterial infections, metabolic engineering intersects with drug discovery, antibiotic resistance mitigation, and the development of novel antimicrobial strategies. This chapter explores cutting-edge approaches that harness microbial metabolism and synthetic biology to address pressing challenges in infectious disease treatment and prevention.

One of the most innovative developments in this field is the use of artificial intelligence to autonomously explore microbial metabolism. The study by Dama et al. introduces BacterAI, a deep reinforcement learning system capable of designing and executing experiments without prior biological knowledge. By converting research questions into game-like formats, BacterAI learns metabolic rules through trial and error, as demonstrated in its analysis of amino acid requirements for *Streptococcus gordonii*. This approach accelerates the discovery of metabolic traits in understudied bacteria and holds promise for expanding our understanding of microbial communities and their therapeutic potential.

The evolution of antibiotic tolerance presents a major obstacle to effective treatment. Zheng et al. investigate how the metabolic dependence of antibiotics influences the trajectory of tolerance evolution. Their study reveals that antibiotics with strong metabolic dependence (SDM) promote faster tolerance development than

V. Soni (✉)
Department of Medicine, Weill Cornell Medicine, New York, NY, USA
e-mail: vis2032@med.cornell.edu

A. Upadhyay
Hospital for Special Surgery, New York, NY, USA

V. Soni, A. Upadhyay (eds.), *Metabolism of Bacterial Infection*,
https://doi.org/10.1007/978-3-031-46895-7_10

those with weak dependence (WDM). By cycling between SDM and WDM antibiotics, it may be possible to delay tolerance evolution and improve treatment outcomes. This insight underscores the importance of considering metabolic context in antibiotic regimen design.

Natural products remain a rich source of antimicrobial agents. Brilisauer et al. identify 7-deoxy-sedoheptulose, a cyanobacterial antimetabolite that inhibits the shikimate pathway—a metabolic route absent in mammals. This compound exhibits herbicidal and antimicrobial activity, offering a blueprint for developing selective therapeutics with minimal side effects. The study highlights the untapped potential of cyanobacteria in producing bioactive metabolites with applications in medicine and agriculture.

Metabolic engineering also enables the biosynthesis of complex antibiotics in tractable hosts. Shomar et al. successfully reconstitute a carbapenem synthesis pathway in *Escherichia coli*, achieving higher yields than native producers. By optimizing gene expression and stabilizing host cells against antibiotic-induced lysis, the researchers demonstrate the feasibility of using *E. coli* as a platform for scalable antibiotic production. This strategy opens avenues for generating novel derivatives and combating resistant pathogens.

Targeting bacterial persisters—dormant cells that evade antibiotics—is another frontier in therapeutic innovation. Allison et al. show that specific metabolites can potentiate aminoglycosides against *E. coli* persisters by enhancing proton-motive force and drug uptake. This metabolite-enabled eradication strategy offers a promising approach to treating chronic infections and biofilms, where persisters contribute to relapse and resistance.

Marine organisms also contribute to the search for new antimicrobials. Hamayeli et al. explore the bioactive metabolites of the sea anemone *Stichodactyla haddoni*, demonstrating antibacterial and antibiofilm activity against resistant strains. The study emphasizes the importance of exploring under-researched marine environments for novel drug discovery.

Optimization of metabolic pathways for industrial applications is exemplified by Shen et al., who develop a kinetic model of the Weimberg pathway for converting xylose to α-ketoglutarate. By integrating experimental data with computational modeling, they identify rate-limiting steps and enhance conversion efficiency. This approach supports the rational design of biocatalytic systems for biofuel and chemical production.

Understanding the metabolic basis of persister formation is critical for developing anti-persistence therapies. Orman et al. reveal that respiration during the stationary phase is essential for persister development in *E. coli*. Inhibiting respiration impairs persister formation, suggesting a metabolic vulnerability that could be targeted to prevent chronic infections.

Garcia and Gardner review bacterial metabolism of α-diglucosides, such as trehalose and kojibiose, highlighting their roles in stress protection and potential applications in biotechnology. Characterizing these pathways may enable the development of probiotics and prebiotics that modulate microbial communities and enhance host health.

Finally, Wang et al. investigate the multi-target action of silver ions against *Staphylococcus aureus*, revealing disruption of glycolysis, oxidative stress response, and energy metabolism. Silver's broad-spectrum activity and reduced resistance potential make it a valuable candidate for combating antibiotic-resistant infections.

Together, these studies showcase the diverse strategies and innovations in metabolic engineering and therapeutic development. This chapter provides a comprehensive overview of how microbial metabolism can be harnessed to address the challenges of bacterial infections and antibiotic resistance.

10.2 Machine Generated Summaries

Disclaimer: The summaries in this chapter were generated from Springer Nature publications using abstractive AI auto-summarization: An abstraction-based summarizer creates new text based on deep learning. New phrases are created to summarize the content. As the constituted sentences are machine selected, they may not fully reflect the body of the work, so we strongly advise that the original content is read and cited. The auto-generated summaries were curated by the editor to meet Springer Nature publication standards.

BacterAI Maps Microbial Metabolism Without Prior Knowledge [89]

This is a machine-generated summary of:

Dama, Adam C.; Kim, Kevin S.; Leyva, Danielle M.; Lunkes, Annamarie P.; Schmid, Noah S.; Jijakli, Kenan; Jensen, Paul A. BacterAI Maps Microbial Metabolism Without Prior Knowledge [89]

Published in: Nature Microbiology (2023)

Link to original: https://doi.org/10.1038/s41564-023-01376-0

If you want to cite the papers, please refer to the original.

For technical reasons we could not place the page where the original quote is coming from.

Research Highlights

The microbiome revolution has unveiled a vast array of bacterial species, many of which remain unstudied due to the sheer volume and complexity of the microbial world. Traditional scientific methods struggle to keep pace with this expanding tree of life, necessitating innovative approaches to accelerate discovery. A significant knowledge gap exists in efficiently studying these lesser-known bacteria, particularly due to the lack of data required for deploying autonomous research agents. Now, a new study by Adam C. Dama and colleagues introduces BacterAI, an AI-driven system that employs deep reinforcement learning (RL) to autonomously design and execute experiments, thereby uncovering the metabolic requirements of understudied bacteria.

BacterAI operates by converting biological research questions into a game-like format, allowing it to apply RL techniques traditionally used in complex games. The system begins with no prior knowledge and learns through trial and error by conducting automated experiments. In their study, Dama et al. tasked BacterAI with determining the amino acid requirements for the growth of *Streptococcus gordonii*, an oral bacterium. The AI agent autonomously selected experiments each day, leveraging a laboratory automation pipeline to perform up to 10,000 liquid-handling operations per assay. BacterAI's neural network model predicted growth outcomes based on experimental data, refining its predictions over time until it could accurately forecast results before conducting new experiments. This iterative process enabled BacterAI to identify specific amino acid combinations that support bacterial growth.

The study demonstrates significant advances in automated biological research by showcasing BacterAI's ability to learn metabolic rules without human bias or preconceptions. Future studies could explore applying BacterAI's methodology to other microbial species or environmental conditions, potentially expanding our understanding of microbial metabolism beyond current genomic databases. Additionally, integrating transfer learning allows BacterAI to leverage prior knowledge from related species or conditions, enhancing its efficiency in discovering metabolic differences among closely related microbes. This approach not only accelerates the study of individual bacterial species but also holds promise for broader applications in microbiome research and beyond.

Modulating the Evolutionary Trajectory of Tolerance Using Antibiotics with Different Metabolic Dependencies [90]

This is a machine-generated summary of:

Zheng, Erica J.; Andrews, Ian W.; Grote, Alexandra T.; Manson, Abigail L.; Alcantar, Miguel A.; Earl, Ashlee M.; Collins, James J. Modulating the Evolutionary Trajectory of Tolerance Using Antibiotics with Different Metabolic Dependencies [90]

Published in: Nature Communications (2022)
Link to original: https://doi.org/10.1038/s41467-022-30272-0

If you want to cite the papers, please refer to the original.

For technical reasons we could not place the page where the original quote is coming from.

Research Highlights

Antibiotic tolerance, a phenomenon where bacteria survive antibiotic treatment without acquiring resistance, is evolving rapidly due to repeated antibiotic exposure. This evolution is driven by heritable mutations that encode tolerance, posing a significant challenge to the efficacy of antibiotics. Most studies on tolerance evolution have focused on antibiotics that depend heavily on bacterial metabolism, which are less effective in conditions where bacterial metabolism is downregulated. Now, a new study by Erica J. Zheng and colleagues investigates how the metabolic dependence of antibiotics influences the evolution of bacterial tolerance.

The researchers conducted parallel evolution experiments using *E. coli* exposed to two types of antibiotics: those with strong dependence on metabolism (SDM) and those with weak dependence (WDM). They found that tolerance evolved more readily with SDM antibiotics than with WDM antibiotics. The study involved daily cycles of antibiotic exposure and recovery, allowing the team to observe changes in bacterial survival over time. Whole-genome sequencing revealed that deletion of the sodium/proton antiporter gene nhaA was a key mutation conferring tolerance during SDM treatment. This mutation led to metabolic suppression, providing a survival advantage under SDM antibiotic pressure but not under WDM conditions.

The study highlights the potential for using cycling strategies involving both SDM and WDM antibiotics to delay the evolution of tolerance while minimizing the use of toxic WDM drugs. Future studies could explore optimizing these cycling strategies to balance efficacy and toxicity concerns in clinical settings. The findings underscore the importance of considering antibiotic metabolism dependence in designing treatment regimens that are more robust against bacterial adaptation strategies, potentially improving outcomes for patients with chronic infections where tolerance is prevalent.

Cyanobacterial Antimetabolite 7-Deoxy-Sedoheptulose Blocks the Shikimate Pathway to Inhibit the Growth of Prototrophic Organisms [91]

This is a machine-generated summary of:

Brilisauer, Klaus; Rapp, Johanna; Rath, Pascal; Schöllhorn, Anna; Bleul, Lisa; Weiß, Elisabeth; Stahl, Mark; Grond, Stephanie; Forchhammer, Karl Cyanobacterial Antimetabolite 7-Deoxy-Sedoheptulose Blocks the Shikimate Pathway to Inhibit the Growth of Prototrophic Organisms [91]

Published in: Nature Communications (2019)

Link to original: https://doi.org/10.1038/s41467-019-08476-8

If you want to cite the papers, please refer to the original.

For technical reasons we could not place the page where the original quote is coming from.

Research Highlights

Cyanobacteria, known for their role as oxygen-producing photoautotrophs, have garnered significant interest in natural product research due to their ability to produce a diverse array of secondary metabolites. These compounds often exhibit antiviral, antibacterial, antifungal, or herbicidal activities, making them promising candidates for applications in health, agriculture, and industry. Despite the broad range of bioactive compounds produced by cyanobacteria, few are classified as classical antimetabolites—compounds that inhibit biological processes by mimicking natural substrates of enzymes. Now, a new study by Klaus Brilisauer and colleagues identifies a novel antimetabolite from *Synechococcus elongatus* PCC 7942 that targets the shikimate pathway in vivo.

The study involved isolating and characterizing a hydrophilic compound from the supernatants of stationary *S. elongatus* cultures. The researchers employed bioactivity-guided isolation techniques, including size-exclusion chromatography and high-performance liquid chromatography (HPLC), to purify the compound.

Structural elucidation using nuclear magnetic resonance (NMR) spectroscopy revealed the compound to be 7-deoxy-sedoheptulose. The team demonstrated that this compound inhibits the enzyme dehydroquinate synthase within the shikimate pathway, leading to the accumulation of its substrate and subsequent metabolic disruption. This inhibition was shown to affect cyanobacterial growth significantly and exhibited herbicidal effects on *Arabidopsis thaliana* seedlings.

The discovery of 7-deoxy-sedoheptulose as a natural inhibitor of the shikimate pathway represents a significant advancement in understanding cyanobacterial secondary metabolites. This finding opens new avenues for developing novel herbicides and antimicrobial agents targeting pathways absent in mammals, thus minimizing potential side effects. Future studies could explore optimizing production methods for this compound and investigating its broader ecological roles and potential applications in agriculture and medicine. The study highlights the untapped potential of simple unicellular cyanobacteria in producing bioactive compounds with significant biotechnological implications.

Metabolic Engineering of a Carbapenem Antibiotic Synthesis Pathway in **Escherichia coli** *[92]*

This is a machine-generated summary of:

Shomar, Helena; Gontier, Sophie; van den Broek, Niels J. F.; Tejeda Mora, Héctor; Noga, Marek J.; Hagedoorn, Peter-Leon; Bokinsky, Gregory Metabolic Engineering of a Carbapenem Antibiotic Synthesis Pathway in *Escherichia coli* [92]

Published in: Nature Chemical Biology (2018)

Link to original: https://doi.org/10.1038/s41589-018-0084-6

If you want to cite the papers, please refer to the original.

For technical reasons we could not place the page where the original quote is coming from.

Research Highlights

Antimicrobial resistance poses a significant threat to global health, necessitating the development of new antibiotics. Traditional methods of antibiotic discovery have often resulted in compounds that are not viable for large-scale production due to low yields. The challenge is further compounded by the inability to cultivate many microbial species in the laboratory, limiting the potential for metabolic engineering. Now, a new study by Helena Shomar and colleagues addresses this issue by engineering *Escherichia coli* as a heterologous host for the biosynthesis of carbapenem antibiotics, achieving higher titers than those produced by native microbial hosts.

The study utilized *E. coli* due to its fast growth and genetic tractability, which allows for rapid design and testing of engineered pathways. The researchers cloned a codon-optimized minimal carbapenem pathway into a high-copy-number plasmid and transformed it into *E. coli*. Cultures were grown in MOPS-based minimal medium supplemented with glucose and glutamate, and an LC-MS method was developed to quantify pathway metabolites. By co-expressing specific genes from *Pectobacterium carotovorum*, they achieved a significant increase in carbapenem productivity. Additionally, they explored strategies to stabilize *E. coli* cells against carbapenem-induced lysis, such as delaying pathway induction and inhibiting fatty acid synthesis using cerulenin or genetic modifications.

The advances made in this study demonstrate the potential of *E. coli* as a platform for producing bactericidal antibiotics like carbapenems at higher titers than previously possible with native hosts. Future studies could focus on optimizing production conditions further and exploring other genetically tractable hosts for antibiotic biosynthesis. The approach also opens avenues for generating novel antibiotic derivatives through combinatorial biochemistry or enzyme engineering, potentially expanding the diversity of available antibiotics to combat resistant bacterial strains effectively.

Metabolite-Enabled Eradication of Bacterial Persisters by Aminoglycosides [93]

This is a machine-generated summary of:

Allison, Kyle R.; Brynildsen, Mark P.; Collins, James J. Metabolite-Enabled Eradication of Bacterial Persisters by Aminoglycosides [93]

Published in: Nature (2011)

Link to original: https://doi.org/10.1038/nature10069

If you want to cite the papers, please refer to the original.

For technical reasons we could not place the page where the original quote is coming from.

Research Highlights

Bacterial persisters are dormant cells that can survive antibiotic treatment, posing a significant challenge in treating chronic infections. Aminoglycosides, which target ribosomes, are typically ineffective against these dormant cells despite ongoing translation. The energy-dependent nature of aminoglycoside uptake suggests that metabolic stimulation could enhance their efficacy against persisters. Now, a new study by Kyle R. Allison and colleagues explores this hypothesis by investigating the role of specific metabolites in potentiating aminoglycosides against *E. coli* persisters.

The researchers conducted a comprehensive screen of metabolites to identify those capable of enhancing the activity of the aminoglycoside gentamicin against *E. coli* persisters. They focused on carbon sources that intersect with glycolysis and other metabolic pathways. Persisters were isolated and treated with gentamicin in the presence of individual metabolites. The study found that metabolites entering upper glycolysis, such as pyruvate, significantly enhanced gentamicin's bactericidal activity, reducing persister viability by three orders of magnitude. This potentiation was linked to increased aminoglycoside uptake facilitated by an elevated proton-motive force (PMF), induced by these metabolites. Further experiments confirmed that this mechanism was specific to aminoglycosides and did not revert persisters to a growth state susceptible to other antibiotics like quinolones or β-lactams.

The study advances our understanding of bacterial persistence by demonstrating that specific metabolic stimuli can prime persisters for elimination by aminoglycosides through PMF induction. This finding suggests potential clinical applications, such as using PMF-stimulating metabolites as adjuvants to enhance aminoglycoside efficacy in treating chronic infections and biofilms. Future studies could explore the applicability of this approach to other bacterial species and infection models, potentially broadening the therapeutic options for persistent bacterial infections. Additionally, understanding the precise biochemical pathways involved may lead to more targeted strategies for combating antibiotic resistance in both Gram-negative and Gram-positive bacteria.

*The Antibacterial and Antibiofilm Activity of Sea Anemone (***Stichodactyla haddoni***) Against Antibiotic-Resistant Bacteria and Characterization of Bioactive Metabolites [94]*

This is a machine-generated summary of:

Hamayeli, Homa; Hassanshahian, Mehdi; Askari Hesni, Majid The Antibacterial and Antibiofilm Activity of Sea Anemone (*Stichodactyla haddoni*) Against Antibiotic-Resistant Bacteria and Characterization of Bioactive Metabolites [94]

Published in: International Aquatic Research (2019)

Link to original: https://doi.org/10.1007/s40071-019-0221-1

For technical reasons we could not place the page where the original quote is coming from.

Research Highlights

The marine environment is a rich source of natural products with unique structural features not found in terrestrial organisms. Marine invertebrates, such as sea anemones, are particularly noted for their bioactive metabolites, which hold potential for novel antimicrobial drug development. Despite the known diversity of microbial communities associated with tropical reef-building corals, the microbiomes of other anthozoans remain largely unexplored. Now, a new study by Homa Hamayeli and colleagues investigates the bioactive metabolites of the sea anemone *Stichodactyla haddoni* from the Persian Gulf, focusing on its antibacterial and antibiofilm activities against pathogenic bacteria.

The study involved collecting *S. haddoni* specimens from Gheshm Islands in the Persian Gulf and extracting bioactive compounds from different parts of the anemone using methanol. The extracts were tested against six antibiotic-resistant bacteria using standard disc diffusion and microtiter plate methods to assess antibacterial activity. The researchers also evaluated the extracts' ability to inhibit and disrupt biofilm formation. Gas chromatography-mass spectrometry (GC-MS) analysis was conducted to identify the chemical composition of the extracts. The results showed that *S. haddoni* extracts exhibited significant antibacterial activity, particularly against *Pseudomonas aeruginosa*, and demonstrated varying degrees of biofilm inhibition and disruption across different bacterial species.

This research marks a significant advance in understanding the antimicrobial potential of marine organisms from the Persian Gulf, highlighting *S. haddoni* as a promising source of bioactive compounds. The study's findings suggest that these compounds could be further explored for their potential applications in preventing bacterial biofilms, which are notoriously resistant to conventional antibiotics. Future studies could focus on isolating specific compounds responsible for these activities and evaluating their efficacy in clinical settings or developing them into new antimicrobial agents. This work underscores the importance of exploring under-researched marine environments for novel drug discovery opportunities.

A Combined Experimental and Modelling Approach for the Weimberg Pathway Optimisation [95]

This is a machine-generated summary of:

Shen, Lu; Kohlhaas, Martha; Enoki, Junichi; Meier, Roland; Schönenberger, Bernhard; Wohlgemuth, Roland; Kourist, Robert; Niemeyer, Felix; van Niekerk, David; Bräsen, Christopher; Niemeyer, Jochen; Snoep, Jacky; Siebers, Bettina A Combined Experimental and Modelling Approach for the Weimberg Pathway Optimisation [95]

Published in: Nature Communications (2020)

Link to original: https://doi.org/10.1038/s41467-020-14830-y

If you want to cite the papers, please refer to the original.

For technical reasons we could not place the page where the original quote is coming from.

Research Highlights

Lignocellulosic biomass is a promising renewable resource for second-generation biofuel and value-added chemical production, but its complex structure poses conversion challenges. The oxidative Weimberg pathway, found in bacteria like *Caulobacter crescentus*, offers a thermodynamically favorable route for converting d-xylose to α-ketoglutarate, a key intermediate in the citric acid cycle. However, existing whole-cell biocatalysis approaches using this pathway have been suboptimal due to issues like intermediate accumulation and poor protein expression. Now, a new study by Lu Shen and colleagues addresses these challenges by developing an iterative multistep approach to construct a quantitative kinetic model of the Weimberg pathway.

The study involved cloning and expressing the five genes of the d-xylose operon from *C. crescentus* in *E. coli*, followed by purification and detailed kinetic characterization of the recombinant enzymes. The researchers developed a model based on initial rate kinetics and validated it using cell-free extract conversion assays. They identified rate-limiting steps and optimized the enzyme cascade for higher conversion efficiency through computational design. The model accurately predicted pathway dynamics under various conditions, including NAD+ recycling and enzyme omission experiments, demonstrating its utility in rational pathway design.

The advances made in this study provide a robust framework for optimizing the Weimberg pathway for industrial applications. The model's ability to predict pathway dynamics in both purified enzyme cascades and cell-free extracts highlights its potential for improving metabolic engineering strategies. Future studies could focus on applying this model to other metabolic pathways or integrating it with whole-cell systems to enhance bioproduct yields. Additionally, the successful demonstration of biotechnological applications, such as producing α-ketoglutarate and other valuable compounds, underscores the industrial relevance of this research.

Inhibition of Stationary Phase Respiration Impairs Persister Formation in E. coli *[96]*

This is a machine-generated summary of:
Orman, Mehmet A.; Brynildsen, Mark P. Inhibition of Stationary Phase Respiration Impairs Persister Formation in *E. coli* [96]
Published in: Nature Communications (2015)
Link to original: https://doi.org/10.1038/ncomms8983

Research Highlights

Persisters are a small subset of bacterial cells that exhibit temporary tolerance to antibiotics, despite being genetically identical to their susceptible counterparts. This tolerance is particularly problematic in biofilm infections, where persisters are shielded from both antibiotics and the host immune system. While anti-persister therapies could improve treatment outcomes for biofilm infections, which are prevalent in US hospitals, strategies to eliminate persisters remain limited due to a lack of understanding of their phenotype. Now, a new study by Mehmet A. Orman and colleagues investigates the role of metabolic activity in the formation of type I persisters in stationary phase cultures of *Escherichia coli*.

The study utilized fluorescence-activated cell sorting (FACS) and a redox-sensitive dye to measure metabolic activity in stationary phase cultures. By segregating the bacterial population into subpopulations based on redox activity, the researchers found that type I persisters were predominantly absent from cells with low redox activity. Genetic and biochemical analyses revealed that inhibiting respiration during the nutrient-depleted stationary phase effectively prevented persister formation. The study also demonstrated that self-digestion processes, such as protein degradation and RNA integrity loss, were associated with persister formation. These findings suggest that respiration is a critical factor in persister development and could be targeted for anti-persister strategies.

The research provides new insights into the metabolic underpinnings of persister cell formation, highlighting respiration as a potential target for therapeutic intervention. Future studies could explore how these findings translate to other bacterial species or conditions and investigate additional metabolic pathways involved in persistence. The identification of respiration as a key player in persister formation opens avenues for developing treatments that could reduce the prevalence of persistent infections by targeting nongrowing bacterial communities.

Bacterial α-Diglucoside Metabolism: Perspectives and Potential for Biotechnology and Biomedicine [97]

This is a machine-generated summary of:

Garcia, Cecelia A.; Gardner, Jeffrey G. Bacterial α-Diglucoside Metabolism: Perspectives and Potential for Biotechnology and Biomedicine [97]

Published in: Applied Microbiology and Biotechnology (2021)

Link to original: https://doi.org/10.1007/s00253-021-11322-x

If you want to cite the papers, please refer to the original.

For technical reasons we could not place the page where the original quote is coming from.

Research Highlights

Bacterial survival is heavily influenced by their ability to acquire and utilize nutrients from their environment, with competition for these resources driving community development and affecting species composition. Within the human gut microbiome, dietary changes can significantly alter bacterial composition, highlighting the importance of understanding nutrient utilization. Of particular interest is the role of rare nutrient sources in disease progression and the potential for therapeutic interventions through probiotics and prebiotics. Now, a new study by Cecelia A. Garcia and colleagues reviews the current understanding of bacterial metabolism of α-diglucosides—specifically trehalose, kojibiose, nigerose, and isomaltose—highlighting critical structural features that affect enzyme activity.

The study delves into the diverse mechanisms bacteria employ to metabolize these uncommon α-diglucosides. Trehalose is primarily used by bacteria as a stress protectant, with its accumulation aiding in survival under various stress conditions. The review discusses trehalose's role as a structural molecule and energy source in certain bacterial species, such as *Mycobacterium* and *Corynebacterium*, which incorporate modified trehalose into their membrane glycolipids. The authors also explore the biosynthetic pathways for trehalose production, noting five potential pathways with varying prevalence among bacteria. For kojibiose, nigerose, and isomaltose, the review highlights recent advances in understanding their biosynthesis and degradation pathways, emphasizing the enzymes involved and their potential industrial applications.

The review concludes by identifying key areas for future research to advance biotechnology and biomedicine applications involving α-diglucosides. Future studies should focus on characterizing the prevalence of bacterial α-diglucoside metabolic enzymes across diverse microbial consortia and understanding their regulation. Additionally, elucidating the transport mechanisms for these sugars within bacterial cells will be crucial for optimizing their use in industrial processes. The authors propose that a comprehensive understanding of both biosynthesis and degradation pathways will be essential for engineering organisms capable of efficiently producing these metabolites for therapeutic and industrial purposes.

Multi-Target Mode of Action of Silver Against Staphylococcus aureus *Endows It with Capability to Combat Antibiotic Resistance [98]*

This is a machine-generated summary of:

Wang, Haibo; Wang, Minji; Xu, Xiaohan; Gao, Peng; Xu, Zeling; Zhang, Qi; Li, Hongyan; Yan, Aixin; Kao, Richard Yi-Tsun; Sun, Hongzhe Multi-Target Mode of Action of Silver Against *Staphylococcus aureus* Endows it with Capability to Combat Antibiotic Resistance [98]

Published in: Nature Communications (2021)

Link to original: https://doi.org/10.1038/s41467-021-23659-y

Research Highlights

Staphylococcus aureus is a leading cause of hospital-acquired infections, with rising cases of antibiotic-resistant strains such as methicillin-resistant *S. aureus*

(MRSA) posing significant treatment challenges. The COVID-19 pandemic has exacerbated this issue due to increased antibiotic use, highlighting the urgent need for alternative antimicrobial strategies. Metal ions, known for their broad-spectrum antimicrobial properties, have gained attention as potential solutions. Despite extensive research on silver ions (Ag+), their precise antimicrobial mechanisms against *S. aureus* remain unclear. Now, a new study by Haibo Wang and colleagues systematically identifies Ag+-binding proteins in *S. aureus*, revealing the bactericidal action of silver at the molecular level.

The study employed a customized approach combining liquid chromatography and gel electrophoresis with inductively coupled plasma mass spectrometry (LC-GE-ICP-MS) to identify Ag+-binding proteins in *S. aureus*. The researchers exposed *S. aureus* cells to silver nitrate and analyzed the proteins that bound to Ag+. They discovered that silver targets multiple biological pathways by disrupting key proteins involved in glycolysis, oxidative stress response, and energy metabolism. The study found that Ag+ primarily binds to proteins involved in glycolysis and oxidative stress defense, leading to reactive oxygen species (ROS) accumulation and bacterial cell death. This multi-target mode of action suggests that silver can effectively combat *S. aureus* by interfering with several critical cellular processes simultaneously.

The findings highlight the potential of silver as a sustainable antimicrobial agent against antibiotic-resistant bacteria like MRSA. By targeting multiple pathways, silver reduces the likelihood of resistance development compared to conventional antibiotics with specific targets. Future study could explore combining silver with antibiotics to enhance efficacy and resensitize resistant strains to treatment. However, challenges remain in improving the biocompatibility of silver for clinical use due to its potential toxicity in humans. Further investigation into the interactions between silver and identified Ag+-binding proteins may provide insights for developing new antibacterial therapies and extending the lifespan of existing antibiotics.

10.3 Conclusion

The studies presented in this chapter feature the transformative potential of metabolic engineering and novel therapeutics in addressing the global challenge of bacterial infections and antimicrobial resistance. By leveraging microbial metabolism, synthetic biology, and systems-level insights, researchers are developing innovative strategies that go beyond traditional antibiotic paradigms, offering new avenues for treatment, prevention, and biotechnological advancement.

Artificial intelligence is revolutionizing microbial research, as demonstrated by BacterAI. This autonomous system learns metabolic rules through reinforcement learning, enabling rapid and unbiased exploration of bacterial physiology. Its success in identifying amino acid requirements for *Streptococcus gordonii* exemplifies how AI can accelerate discovery in microbiome science and therapeutic development, especially for understudied or unculturable species.

Understanding the metabolic dependencies of antibiotics is crucial for designing robust treatment regimens. Zheng et al. show that antibiotics with strong metabolic dependence promote faster evolution of tolerance, while weakly dependent drugs delay this process. Strategic cycling between these drug types could mitigate tolerance development, enhancing long-term efficacy and reducing the risk of chronic infections.

Natural antimetabolites offer selective and potent antimicrobial activity. The discovery of 7-deoxy-sedoheptulose by Brilisauer et al. reveals a cyanobacterial compound that inhibits the shikimate pathway, absent in mammals. This specificity minimizes side effects and positions such compounds as promising candidates for herbicides and antimicrobials. The study underscores the value of exploring microbial secondary metabolites for novel therapeutic agents.

Metabolic engineering enables scalable antibiotic production in genetically tractable hosts. Shomar et al. reconstitute a carbapenem biosynthesis pathway in *E. coli*, achieving higher yields than native producers. Their approach demonstrates the feasibility of using engineered microbes for industrial-scale antibiotic synthesis and opens the door to generating novel derivatives through enzyme engineering and combinatorial biosynthesis.

Targeting bacterial persisters is a major therapeutic challenge. Allison et al. reveal that specific metabolites can potentiate aminoglycosides against dormant *E. coli* cells by enhancing proton-motive force and drug uptake. This metabolite-enabled strategy offers a new approach to eradicating persistent infections and biofilms, which are often refractory to conventional antibiotics.

Marine ecosystems are rich sources of bioactive compounds. Hamayeli et al. identify antibacterial and antibiofilm metabolites from the sea anemone *Stichodactyla haddoni*, demonstrating activity against resistant pathogens. Their findings highlight the untapped potential of marine biodiversity in drug discovery and the importance of exploring under-researched environments.

Optimization of metabolic pathways for industrial applications is exemplified by Shen et al., who develop a kinetic model of the Weimberg pathway for converting xylose to α-ketoglutarate. Their iterative modeling and experimental validation enhance conversion efficiency, supporting the rational design of biocatalytic systems for biofuel and chemical production.

Respiration plays a critical role in persister formation. Orman et al. show that inhibiting respiration during the stationary phase impairs the development of type I persisters in *E. coli*. This insight identifies a metabolic vulnerability that could be exploited to prevent chronic infections and improve treatment outcomes.

Garcia and Gardner's review of α-diglucoside metabolism highlights the importance of rare sugars in bacterial physiology and their potential applications in biotechnology and biomedicine. Understanding these pathways may inform the development of targeted probiotics and prebiotics that modulate microbial communities and support host health.

Silver ions offer a multi-target antimicrobial strategy. Wang et al. demonstrate that silver disrupts glycolysis, oxidative stress response, and energy metabolism in *Staphylococcus aureus*, leading to cell death. This broad-spectrum activity reduces the likelihood of resistance development and positions silver as a valuable adjunct or alternative to conventional antibiotics.

In conclusion, metabolic engineering and novel therapeutics represent a frontier in microbiology and medicine. The studies in this chapter showcase diverse approaches—from AI-driven discovery and synthetic biology to natural product exploration and metabolic pathway optimization—that are reshaping our ability to understand and treat bacterial infections. As these technologies continue to evolve, they will play a pivotal role in overcoming antibiotic resistance, enhancing microbial control, and advancing global health.

Bibliography

1. Schrittwieser, J. et al. Mastering Atari, Go, chess and shogi by planning with a learned model. *Nature 588*, 604–609 (2020).
2. Silver, D. et al. A general reinforcement learning algorithm that masters chess, dhogi, and Go through self-play. *Science 362*, 1140–1144 (2018).
3. Fridman, O., Goldberg, A., Ronin, I., Shoresh, N. & Balaban, N. Q. Optimization of lag time underlies antibiotic tolerance in evolved bacterial populations. *Nature 513*, 418–421 (2014).
4. Levin-Reisman, I. et al. Antibiotic tolerance facilitates the evolution of resistance. *Science 355*, 827–830 (2017).
5. Sulaiman, J. E. & Lam, H. Proteomic investigation of tolerant *Escherichia coli* populations from cyclic antibiotic treatment. *J. Proteome Res. 19*, 900–913 (2020).
6. Sulaiman, J. E. & Lam, H. Proteomic study of the survival and resuscitation mechanisms of filamentous persisters in an evolved *Escherichia coli* population from cyclic ampicillin treatment. *mSystems 5*, e00462–20 (2020).
7. Khare, A. & Tavazoie, S. Extreme antibiotic persistence via heterogeneity-generating mutations targeting translation. *mSystems 5*, e00847–19 (2020).
8. Zheng, E. J., Stokes, J. M. & Collins, J. J. Eradicating bacterial persisters with combinations of strongly and weakly metabolism-dependent antibiotics. *Cell. Chem. Biol. 27*, 1–9 (2020).
9. Lopatkin, A. J. et al. Bacterial metabolic state more accurately predicts antibiotic lethality than growth rate. *Nat. Microbiol. 19*, 1–9 (2019).
10. Stokes, J. M., Lopatkin, A. J., Lobritz, M. A. & Collins, J. J. Bacterial metabolism and antibiotic efficacy. *Cell Metab. 30*, 251–259 (2019).
11. Cornforth, D. M. et al. *Pseudomonas aeruginosa* transcriptome during human infection. *Proc. Natl Acad. Sci. USA 115*, E5125–E5134 (2018).
12. Grant SS, Hung DT (2013) Persistent bacterial infections, antibiotic tolerance, and the oxidative stress response. Virulence 4:273–283. https://doi.org/10.4161/viru.23987
13. Meylan, S., Andrews, I. W. & Collins, J. J. Targeting antibiotic tolerance, pathogen by pathogen. *Cell 172*, 1228–1238 (2018).
14. Høiby, N., Ciofu, O. & Bjarnsholt, T. *Pseudomonas aeruginosa* biofilms in cystic fibrosis. *Future Microbiol. 5*, 1663–1674 (2010).
15. Kordes, A. et al. Genetically diverse *Pseudomonas aeruginosa* populations display similar transcriptomic profiles in a cystic fibrosis explanted lung. *Nat. Commun. 10*, 3397 (2019).
16. Bjarnsholt, T. et al. The importance of understanding the infectious microenvironment. *Lancet Infect Dis*. https://doi.org/10.1016/S1473-3099(21)00122-5 (2021).
17. Prax, M. & Bertram, R. Metabolic aspects of bacterial persisters. *Front. Cell. Infect. Microbol. 4*, 148 (2014).
18. Schrader, S. M. et al. Multiform antimicrobial resistance from a metabolic mutation. *Sci. Adv. 7*, eabh2037 (2021).

19. Lopatkin, A. J. et al. Clinically relevant mutations in core metabolic genes confer antibiotic resistance. *Science 371*, eaba0862 (2021).
20. Duke, S. O. & Powles, S. B. Glyphosate: a once-in-a-century herbicide. *Pest. Manag. Sci. 64*, 319–325 (2008).
21. Motta, E. V. S., Raymann, K., Moran, N. A. Glyphosate perturbs the gut microbiota of honey bees. *Proc. Natl Acad. Sci. 115*, 10305–10310 (2018).
22. Lonhienne, T. et al. Structural insights into the mechanism of inhibition of AHAS by herbicides. *Proc. Natl Acad. Sci. 115*, E1945–E1954 (2018).
23. Daniel, G. M., Jian-Guo, W., Thierry, L. & William, G. L. Crystal structure of plant acetohydroxyacid synthase, the target for several commercial herbicides. *FEBS J. 284*, 2037–2051 (2017).
24. Klotz, A. et al. Awakening of a Dormant Cyanobacterium from Nitrogen Chlorosis Reveals a Genetically Determined Program. *Curr. Biol. 26*, 2862–2872 (2016).
25. Medema, M. H., Breitling, R., Bovenberg, R. & Takano, E. Exploiting plug-and-play synthetic biology for drug discovery and production in microorganisms. *Nat. Rev. Microbiol. 9*, 131–137 (2011).
26. Awan, A. R. et al. Biosynthesis of the antibiotic nonribosomal peptide penicillin in baker's yeast. *Nat. Commun. 8*, 15202 (2017).
27. Olano, C., Lombó, F., Méndez, C. & Salas, J. A. Improving production of bioactive secondary metabolites in actinomycetes by metabolic engineering. *Metab. Eng. 10*, 281–292 (2008).
28. McGowan, S. J. et al. Analysis of bacterial carbapenem antibiotic production genes reveals a novel beta-lactam biosynthesis pathway. *Mol. Microbiol. 22*, 415–426 (1996).
29. McGowan, S. J. et al. Analysis of the carbapenem gene cluster of *Erwinia carotovora*: definition of the antibiotic biosynthetic genes and evidence for a novel β-lactam resistance mechanism. *Mol. Microbiol. 26*, 545–556 (1997).
30. Marous, D. R. et al. Consecutive radical *S*-adenosylmethionine methylations form the ethyl side chain in thienamycin biosynthesis. *Proc. Natl Acad. Sci. USA 112*, 10354–10358 (2015).
31. Tuomanen, E. Phenotypic tolerance: the search for β-lactam antibiotics that kill nongrowing bacteria. *Rev. Infect. Dis. 8* (Suppl. 3), S279–S291 (1986).
32. Neidhardt, F. C., Bloch, P. L. & Smith, D. F. Culture medium for enterobacteria. *J. Bacteriol. 119*, 736–747 (1974).
33. Bryan, L. E. & Kwan, S. Roles of ribosomal binding, membrane potential, and electron transport in bacterial uptake of streptomycin and gentamicin. *Antimicrob. Agents Chemother. 23*, 835–845 (1983)
34. Keren, I., Kaldalu, N., Spoering, A., Wang, Y. & Lewis, K. Persister cells and tolerance to antimicrobials. *FEMS Microbiol. Lett. 230*, 13–18 (2004)
35. Hansen, S., Lewis, K. & Vulic, M. Role of global regulators and nucleotide metabolism in antibiotic tolerance in *Escherichia coli*. *Antimicrob. Agents Chemother. 52*, 2718–2726 (2008)
36. Sandoval, R., Leiser, J. & Molitoris, B. A. Aminoglycoside antibiotics traffic to the Golgi complex in LLC-PK1 cells. *J. Am. Soc. Nephrol. 9*, 167–174 (1998)
37. Prakash WG, Babu S, Ravikumar S, Kathiresan K, Arul Prathap S, Chinnapparaj S, Marian MP, Liakath Alikhan S (2007) Antimicrobial activity of tissue and associated bacteria from benthic sea anemone *Stichodactyla haddoni* against microbial pathogens. J Environ Biol 28(4):789–793
38. Mohammadi M, Masoumipour F, Hassanshahian M, Jafarinasab T (2019) Study the antibacterial and antibiofilm activity of *Carum copticum* against antibiotic-resistant bacteria in planktonic and biofilm forms. Microb Pathog 129:99–105. https://doi.org/10.1016/j.micpath.2019.02.002
39. Mohsenipour Z, Hassanshahian M (2015) The inhibitory effect of *Thymus vulgaris* extracts on the planktonic form and biofilm structures of six human pathogenic bacteria. Avicen J Phytomed 5(4):309–317
40. Williams JM, Barnhofer T, Crane C, Herman D, Raes F, Watkins E, Dalgleish T (2007) Autobiographical memory specificity and emotional disorder. Psychol Bull 133(1):122–48

41. Gunasundari V, Ajith Kumar TT, Kumaresan S, Balagurunathan R, Balasubramanian T (2013) Isolation of aliphatic-antibiotic compounds from marine invertebrate, *Heteractis magnifica* "Quoy & Gaimard 1833" against captive marine ornamental fish pathogens. Indian J Geomar Sci 42:807–811
42. Maisonneuve, E., Castro-Camargo, M. & Gerdes, K. (p)ppGpp controls bacterial persistence by stochastic induction of toxin-antitoxin activity. *Cell 154*, 1140–1150 (2013).
43. Balaban, N. Q., Merrin, J., Chait, R., Kowalik, L. & Leibler, S. Bacterial persistence as a phenotypic switch. *Science 305*, 1622–1625 (2004)
44. Gefen, O., Gabay, C., Mumcuoglu, M., Engel, G. & Balaban, N. Q. Single-cell protein induction dynamics reveals a period of vulnerability to antibiotics in persister bacteria. *Proc. Natl Acad. Sci. USA 105*, 6145–6149 (2008)
45. Pearl, S., Gabay, C., Kishony, R., Oppenheim, A. & Balaban, N. Q. Nongenetic Individuality in the Host–Phage Interaction. *PLoS Biol. 6*, e120 (2008).
46. Orman, M. A. & Brynildsen, M. P. Dormancy is not necessary or sufficient for bacterial persistence. *Antimicrob. Agents Chemother. 57*, 3230–3239 (2013).
47. Amato, Stephanie M., Orman, Mehmet A. & Brynildsen, Mark P. Metabolic Control of Persister Formation in Escherichia coli. *Mol. Cell. 50*, 475–487 (2013).
48. Maisonneuve, E., Shakespeare, L. J., Jørgensen, M. G. & Gerdes, K. Bacterial persistence by RNA endonucleases. *Proc. Natl Acad. Sci. USA 108*, 13206–13211 (2011).
49. Amato, S. M. & Brynildsen, M. P. Nutrient transitions are a source of persisters in *Escherichia coli* biofilms. *PLoS ONE 9*, e93110 (2014).
50. Peralta-Yahya PP, Zhang F, Del Cardayre SB, Keasling JD (2012) Microbial engineering for the production of advanced biofuels. Nature 488:320–328. https://doi.org/10.1038/nature11478
51. Elbein AD, Pan YT, Pastuszak I, Carroll D (2003) New insights on trehalose: a multifunctional molecule. Glycobiology 13:17–27. https://doi.org/10.1093/glycob/cwg047
52. Avonce N, Mendoza-Vargas A, Morett E, Iturriag G (2006) Insights on the evolution of trehalose biosynthesis. BMC Evol Biol 6:1–15. https://doi.org/10.1186/1471-2148-6-109
53. Ruhal R, Kataria R, Choudhury B (2013) Trends in bacterial trehalose metabolism and significant nodes of metabolic pathway in the direction of trehalose accumulation. Microb Biotechnol 6:493–502. https://doi.org/10.1111/1751-7915.12029
54. Wannet WJB, Op den Camp HJM, Wisselink HW, van der Drift C, Van Griensven LJLD, Vogels GD (1998) Purification and characterization of trehalose phosphorylase from the commercial mushroom *Agaricus bisporus*. Biochem Biophys Acta 1425:177–188
55. Qu Q, Lee SJ, Boos W (2004) TreT, a novel trehalose glycosyltransferring synthase of the hyperthermophilic archaeon *Thermococcus litoralis*. J Biol Chem 279:47890–47897. https://doi.org/10.1074/jbc.M404955200
56. Nobre A, Alarico S, Fernandes C, Empadinhas N, Da Costa MS (2008) A unique combination of genetic systems for the synthesis of trehalose in *Rubrobacter xylanophilus*: properties of a rare actinobacterial TreT. J Bacteriol 190:7939–7946. https://doi.org/10.1128/JB.01055-08
57. Lombard V, Ramulu HG, Drula E, Coutinho PM, Henrissat B. The carbohydrate-active enzymes database (CAZy) in 2013. Nucleic Acids Res. 2014;42(D1):D490–5. https://doi.org/10.1093/nar/gkt1178.
58. Gibson RP, Gloster TM, Roberts S, Warren RAJ, Storch De Gracia I, García Á, Chiara JL, Davies GJ (2007) Molecular basis for trehalase inhibition revealed by the structure of trehalase in complex with potent inhibitors. Angew Chemie - Int Ed 46:4115–4119. https://doi.org/10.1002/anie.200604825
59. Garcia CA, Narrett JA, Gardner JG (2020) Trehalose degradation in *Cellvibrio japonicus* exhibits no functional redundancy and is solely dependent on the Tre37A enzyme. Appl Environ Microbiol 86:1–15. https://doi.org/10.1128/aem.01639-20
60. Yamamoto T, Yamashita H, Mukai K, Watanabe H, Kubota M, Chaen H, Fukuda S (2006) Construction and characterization of chimeric enzymes of kojibiose phosphorylase and trehalose phosphorylase from *Thermoanaerobacter brockii*. Carbohydr Res 341:2350–2359. https://doi.org/10.1016/j.carres.2006.06.024

61. Okada S, Yamamoto T, Watanabe H, Nishimoto T, Chaen H, Fukuda S, Wakagi T, Fushinobu S (2014) Structural and mutational analysis of substrate recognition in kojibiose phosphorylase. FEBS J 281:778–786. https://doi.org/10.1111/febs.12622
62. Nihira T, Nakai H, Chiku K, Kitaoka M (2012) Discovery of nigerose phosphorylase from *Clostridium phytofermentans*. Appl Microbiol Biotechnol 93:1513–1522. https://doi.org/10.1007/s00253-011-3515-9
63. Nihira T, Miyajima F, Chiku K, Nishimoto M, Kitaoka M, Ohtsubo K, Nakai H (2014) One pot enzymatic production of nigerose from common sugar resources employing nigerose phosphorylase. J Appl Glycosci 61:75–80. https://doi.org/10.5458/jag.jag.jag-2013_012
64. Kitao S, Horiuchi T, Yoshida S, Kusakabe I, Sekine H (1994) Formation of kojibiose and nigerose by sucrose phosphorylase. Biosci Biotechnol Biochem 58:790–791. https://doi.org/10.1271/bbb.58.790
65. Kurakata Y, Uechi A, Yoshida H, Kamitori S, Sakano Y, Nishikawa A, Tonozuka T (2008) Structural insights into the substrate specificity and function of *Escherichia coli* K12 YgjK, a glucosidase belonging to the Glycoside Hydrolase Family 63. J Mol Biol 381:116–128
66. Kang MS, Okuyama M, Mori H, Kimura A (2009) The first α-1,3-glucosidase from bacterial origin belonging to glycoside hydrolase family 31. Biochimie 91:1434–1442. https://doi.org/10.1016/j.biochi.2009.07.018
67. Sanz ML, Gibson GR, Rastall RA (2005) Influence of disaccharide structure on prebiotic selectivity in vitro. J Agric Food Chem 53:5192–5199. https://doi.org/10.1021/jf050276w
68. Harrion MD, Purdue JC, Hoffman AJ (2007) Food products comprising a slowly digestible or digestion resistant carbohydrate composition. U.S. paten
69. Yen CH, Tseng YH, Kuo YW, Lee MC, Chen HL (2011) Long-term supplementation of isomalto-oligosaccharides improved colonic microflora profile, bowel function, and blood cholesterol levels in constipated elderly people—a placebo-controlled, diet-controlled trial. Nutrition 27:445–450. https://doi.org/10.1016/j.nut.2010.05.012
70. Abe K, Kuroda A, Takeshita R (2017) Engineering of *Escherichia coli* to facilitate efficient utilization of isomaltose and panose in industrial glucose feedstock. Appl Microbiol Biotechnol 101:2057–2066. https://doi.org/10.1007/s00253-016-8037-z
71. Luo D, Deng T, Yuan W, Deng H, Jin M (2017) Plasma metabolomic study in Chinese patients with wet age-related macular degeneration. BMC Ophthalmol 17:1–9. https://doi.org/10.1186/s12886-017-0555-7
72. Fatoki T, Sanni D, Momodu D, Ugboko H, Adeseko C, Faleye B (2018) Evaluation of empirical functions and fate of isomaltose. J Appl Life Sci Int 16:1–10. https://doi.org/10.9734/jalsi/2018/39370
73. Schönert S, Buder T, Dahl MK (1998) Identification and enzymatic characterization of the maltose-inducible α-glucosidase MalL (sucrase-isomaltase-maltase) of *Bacillus subtilis*. J Bacteriol 180:2574–2578. https://doi.org/10.1128/jb.180.9.2574-2578.1998
74. Schönert S, Buder T, Dahl MK (1999) Properties of maltose-inducible α-glucosidase MalL (sucrase-isomaltase-maltase) in *Bacillus subtilis*: evidence for its contribution to maltodextrin utilization. Res Microbiol 150:167–177. https://doi.org/10.1016/S0923-2508(99)80033-3
75. Pokusaeva K, O'Connell-Motherway M, Zomer A, Fitzgerald GF, Van Sinderen D (2009) Characterization of two novel α-glucosidases from *Bifidobacterium breve* UCC2003. Appl Environ Microbiol 75:1135–1143. https://doi.org/10.1128/AEM.02391-08
76. Suzuki Y, Tomura Y (1986) Purification and characterization of *Bacillus coagulans* oligo-1,6-glucosidase. Eur J Biochem 158:77–83. https://doi.org/10.1111/j.1432-1033.1986.tb09723.x
77. Garcia CA, Narrett JA, Gardner JG (2019) Complete genome sequences of *Cellvibrio japonicus* strains with improved growth when using α-diglucosides. Microbiol Resour Announc 8:3–5. https://doi.org/10.1128/mra.01077-19
78. Maple, P. A. C., Hamilton-Miller, J. M. T. & Brumfitt, W. World-wide antibiotic resistance in methicillin-resistant *Staphylococcus aureus*. *Lancet 333*, 537–540 (1989).
79. Sakoulas, G. & Moellering, R. C. Jr. Increasing antibiotic resistance among methicillin-resistant *Staphylococcus aureus* strains. *Clin. Infect. Dis. 46*, S360–S367 (2008).

80. Hiramatsu, K. Vancomycin-resistant *Staphylococcus aureus*: a new model of antibiotic resistance. *Lancet Infect. Dis. 1*, 147–155 (2001).
81. Editorial: Antimicrobial resistance in the age of COVID-19. *Nat. Microbiol. 5*, 779 (2020).
82. Rawson, T. M., Ming, D., Ahmad, R., Moore, L. S. P. & Holmes, A. H. Antimicrobial use, drug-resistant infections, and COVID-19. *Nat. Rev. Microbiol. 18*, 409–410 (2020).
83. Wang, H., Zhou, Y., Xu, X., Li, H. & Sun, H. Metalloproteomics in conjunction with other omics for uncovering the mechanism of action of metallodrugs: mechanism-driven new therapy development. *Curr. Opin. Chem. Biol. 55*, 171–179 (2020).
84. Waldron, K. J. & Robinson, N. J. How do bacterial cells ensure that metalloproteins get the correct metal? *Nat. Rev. Microbiol. 7*, 25–35 (2009).
85. Waldron, K. J., Rutherford, J. C., Ford, D. & Robinson, N. J. Metalloproteins and metal sensing. *Nature 460*, 823–830 (2009).
86. Wang, H. et al. Deciphering molecular mechanism of silver by integrated omic approaches enables enhancing its antimicrobial efficacy in *E. coli*. *PLoS. Biol. 17*, e3000292 (2019).
87. Lok, C.-N. et al. Proteomic analysis of the mode of antibacterial action of silver nanoparticles. *J. Proteome Res. 5*, 916–924 (2006).
88. Gordon, O. et al. Silver coordination polymers for prevention of implant infection: thiol interaction, impact on respiratory chain enzymes, and hydroxyl radical induction. *Antimicrob. Agents Chemother. 54*, 4208–4218 (2010).
89. Dama, Adam C.; Kim, Kevin S.; Leyva, Danielle M.; Lunkes, Annamarie P.; Schmid, Noah S.; Jijakli, Kenan; Jensen, Paul A. BacterAI maps microbial metabolism without prior knowledge. Nature Microbiology (2023). https://doi.org/10.1038/s41564-023-01376-0
90. Zheng, Erica J.; Andrews, Ian W.; Grote, Alexandra T.; Manson, Abigail L.; Alcantar, Miguel A.; Earl, Ashlee M.; Collins, James J. Modulating the evolutionary trajectory of tolerance using antibiotics with different metabolic dependencies. Nature Communications (2022). https://doi.org/10.1038/s41467-022-30272-0
91. Brilisauer, Klaus; Rapp, Johanna; Rath, Pascal; Schöllhorn, Anna; Bleul, Lisa; Weiß, Elisabeth; Stahl, Mark; Grond, Stephanie; Forchhammer, Karl. Cyanobacterial antimetabolite 7-deoxysedoheptulose blocks the shikimate pathway to inhibit the growth of prototrophic organisms. Nature Communications (2019). https://doi.org/10.1038/s41467-019-08476-8
92. Shomar, Helena; Gontier, Sophie; van den Broek, Niels J. F.; Tejeda Mora, Héctor; Noga, Marek J.; Hagedoorn, Peter-Leon; Bokinsky, Gregory. Metabolic engineering of a carbapenem antibiotic synthesis pathway in Escherichia coli. Nature Chemical Biology (2018). https://doi.org/10.1038/s41589-018-0084-6
93. Allison, Kyle R.; Brynildsen, Mark P.; Collins, James J. Metabolite-enabled eradication of bacterial persisters by aminoglycosides. Nature (2011). https://doi.org/10.1038/nature10069
94. Hamayeli, Homa; Hassanshahian, Mehdi; Askari Hesni, Majid. The antibacterial and antibiofilm activity of sea anemone (Stichodactyla haddoni) against antibiotic-resistant bacteria and characterization of bioactive metabolites. International Aquatic Research (2019). https://doi.org/10.1007/s40071-019-0221-1
95. Shen, Lu; Kohlhaas, Martha; Enoki, Junichi; Meier, Roland; Schönenberger, Bernhard; Wohlgemuth, Roland; Kourist, Robert; Niemeyer, Felix; van Niekerk, David; Bräsen, Christopher; Niemeyer, Jochen; Snoep, Jacky; Siebers, Bettina. A combined experimental and modelling approach for the Weimberg pathway optimisation. Nature Communications (2020). https://doi.org/10.1038/s41467-020-14830-y
96. Orman, Mehmet A.; Brynildsen, Mark P. Inhibition of stationary phase respiration impairs persister formation in E. coli. Nature Communications (2015). https://doi.org/10.1038/ncomms8983
97. Garcia, Cecelia A.; Gardner, Jeffrey G. Bacterial α-diglucoside metabolism: perspectives and potential for biotechnology and biomedicine. Applied Microbiology and Biotechnology (2021). https://doi.org/10.1007/s00253-021-11322-x
98. Wang, Haibo; Wang, Minji; Xu, Xiaohan; Gao, Peng; Xu, Zeling; Zhang, Qi; Li, Hongyan; Yan, Aixin; Kao, Richard Yi-Tsun; Sun, Hongzhe. Multi-target mode of action of silver against Staphylococcus aureus endows it with capability to combat antibiotic resistance. Nature Communications (2021). https://doi.org/10.1038/s41467-021-23659-y

MIX
Papier aus verantwortungsvollen Quellen
Paper from responsible sources
FSC® C105338

FSC
www.fsc.org

If you have any concerns about our products,
you can contact us on
ProductSafety@springernature.com

In case Publisher is established outside the EU,
the EU authorized representative is:
Springer Nature Customer Service Center GmbH
Europaplatz 3, 69115 Heidelberg, Germany

Printed by Libri Plureos GmbH
in Hamburg, Germany